7/88

3000 800012 51555
St. Louis Community College

WITHDRAWN

D1263921

St. Louis Community College

Library

5801 Wilson Avenue
St. Louis, Missouri 63110

St. Louis Community College
at Meramec
Library

Silent Spring Revisited

Silent Spring
Revisited

Edited by   Gino J. Marco
Robert M. Hollingworth
and
William Durham

AMERICAN CHEMICAL SOCIETY
Washington, DC   1987

Library of Congress Cataloging-in-Publication Data

Silent spring revisited.

Based on a symposium on the topics posed in Rachel Carson's Silent spring, held in Philadelphia, Aug. 1984.

Includes bibliographies and index.

1. Pesticides—Environmental aspects—United States—Congresses.
2. Carson, Rachel, 1907–1964. Silent Spring—Congresses.

I. Marco, Gino J., 1924–   . II. Hollingworth, Robert M., 1939–   .
III. Durham, William, 1922–   . IV. Carson, Rachel, 1907–1964. Silent spring.

[DNLM: 1. Environmental Pollutants—congresses. 2. Insect Control—congresses. 3. Pesticides—congresses. QX 600 S582 1964]

QH545.P4S55   1987   363.7'384   86–25952
ISBN 0-8412-0980-4
ISBN 0-8412-0981-2 (pbk.)

Copyright © 1987

American Chemical Society

All Rights Reserved. The appearance of the code at the bottom of the first page of each chapter in this volume indicates the copyright owner's consent that reprographic copies of the chapter may be made for personal or internal use or for the personal or internal use of specific clients. This consent is given on the condition, however, that the copier pay the stated per copy fee through the Copyright Clearance Center, Inc., 27 Congress Street, Salem, MA 01970, for copying beyond that permitted by Sections 107 or 108 of the U.S. Copyright Law. This consent does not extend to copying or transmission by any means—graphic or electronic—for any other purpose, such as for general distribution, for advertising or promotional purposes, for creating a new collective work, for resale, or for information storage and retrieval systems. The copying fee for each chapter is indicated in the code at the bottom of the first page of the chapter.

The citation of trade names and/or names of manufacturers in this publication is not to be construed as an endorsement or as approval by ACS of the commercial products or services referenced herein; nor should the mere reference herein to any drawing, specification, chemical process, or other data be regarded as a license or as a conveyance of any right or permission, to the holder, reader, or any other person or corporation, to manufacture, reproduce, use, or sell any patented invention or copyrighted work that may in any way be related thereto. Registered names, trademarks, etc., used in this publication, even without specific indication thereof, are not to be considered unprotected by law.

PRINTED IN THE UNITED STATES OF AMERICA

Contents

Specific Environmental Effects

In Conclusion

Appendix and Indexes

Contributors

Briggs, Shirley A.   page 3
Rachel Carson Council, Inc.
8940 Jones Mill Road
Chevy Chase, MD 20815

Carsel, Robert F.   page 71
U.S. Environmental Protection Agency
Environmental Research Laboratory
Athens, GA 30613

Coppage, D. L.   page 49
U.S. Environmental Protection Agency
Office of Pesticide Programs
Washington, DC 20460

Davies, J. E.   page 113
University of Miami School of Medicine
Department of Epidemiology and Public Health
Miami, FL 33177

Doon, R.   page 113
Occupational Health Unit
Ministry of Health and Environment
Port-of-Spain, Trinidad-Tobago
West Indies

Durham, William   page 191
U.S. Environmental Protection Agency
Research Triangle Park, NC 27711

Freed, Virgil H.   page 145
Oregon State University
Agricultural Chemistry Department
Corvallis, OR 97331

Gretch, Fred M.   page 127
Food and Drug Administration
New York Regional Laboratory
850 Third Avenue
Brooklyn, NY 11232

Hall, Russell, J.   page 85
U.S. Department of the Interior
Fish and Wildlife Service
Patuxent Wildlife Research Center
Laurel, MD 20708

Hansen, D. J.   page 49
U.S. Environmental Protection Agency
Environmental Research Laboratory
Narragansett, RI 02882

Hollingworth, Robert M.   page 191
Department of Entomology
Purdue University
West Lafayette, IN 47907

Kohn, Gustave K.   page 159
Zoecon Corporation
Palo Alto, CA 94304

Marco, Gino J.   pages xvii, 191
Ciba-Geigy Corporation
Greensboro, NC 27419

Moore, John A.   page 15
U.S. Environmental Protection Agency
Office of Pesticides and Toxic Substances
Washington, DC 20460

Nimmo, D. R.   page 49
U.S. Environmental Protection Agency
One Denver Place—999 18th Street—Suite 1300
Denver, CO 80202
and
Colorado State University
Department of Fishery and Wildlife Biology
Fort Collins, CO 80523

Pickering, Q. L.   page 49
U.S. Environmental Protection Agency
Office of Research and Development
Cincinnati, OH 45268

Pimentel, David   page 175
Cornell University
Department of Entomology
Ithaca, NY 14853

Rosen, Joseph D.   page 127
Rutgers University
Department of Food Science
New Brunswick, NJ 08903

Smith, Charles N.   page 71
U.S. Environmental Protection Agency
Environmental Research Laboratory
Athens, GA 30613

Wilkinson, C. F.   page 25
Cornell University
Institute for Comparative and Environmental Toxicology
Ithaca, NY 14853
and
Institute for Health Policy Analysis
Georgetown University Medical Center
Washington, DC 20007

Preface

Several years ago, the Pesticide Subcommittee of the Committee on Environmental Improvement wished to sponsor a symposium addressing environmental and health issues concerning pesticides. To this end, as a member of the subcommittee, I (G. J. M.), with the help of my colleagues (R. M. H. and W. D.), chose to address issues accentuated in a classic environmental book of our times, Rachel Carson's *Silent Spring*.

Some 20 years after Rachel Carson wrote her book, it seemed appropriate to place her views in perspective. When published, the book led to much discussion and controversy among scientists and the general public. One result was Congressional action leading to the establishment of the Environmental Protection Agency.

The symposium upon which this book is based was organized to address the issues that Rachel Carson raised and focus on their pertinence for the past, present, and future. The topics in this book reflect the symposium and were selected to help assess how right Rachel Carson was, whether the issues posed in *Silent Spring* are still of major concern, and if the problems have been brought under control. We hoped to answer the question, "Is the present awareness of society such that there would be no future incentive to write another *Silent Spring*?"

We were fortunate to obtain authors who are very knowledgeable, experienced, and noted in their respective fields to address these issues and provide various points of view. We thank them for their contributions and their patience. We also thank the Pesticide Subcommittee and the Pesticide Chemical (now Agrochemicals) Division for their support for the symposium.

GINO J. MARCO
ROBERT M. HOLLINGWORTH
WILLIAM DURHAM

A Summary of *Silent Spring*

Gino J. Marco

Because copies of Rachel Carson's book are no longer readily available, it seems appropriate to summarize what *Silent Spring* emphasized. The following is an attempt to provide a nonjudgmental, brief summary of *Silent Spring*.

In the first several chapters, Rachel Carson stated that the large number of chemicals (approximately 500, many were pesticides) introduced each year was possibly making the earth unfit for all life. Insecticides were becoming deadlier and deadlier. Specialists were concerned only about efficacy and were losing sight of the overall picture. Before World War II, inorganic chemicals were the main pest controls. Arsenicals were greatly used, and toxicological problems occurred. Carson emphasized chlorinated hydrocarbons and organophosphates as the main problems leading to bird and fish kills, human nervous system disorders, and deaths. She noted that herbicides were at one time considered no problem to animals. She explored the possibility of surface and ground water contamination and pointed to leaching, runoff, and direct spray as contamination problems. She explained that water treatment plants did not remove chemicals because multiple chemicals in catch basins could interact to form toxic compounds, and thus cancer hazards from polluted waters would increase in the future.

Carson stated that chemical treatment of soils led to the destruction of beneficial biological species, and that such destruction resulted in imbalance to the ecosystem. Also, wildlife that ate chemically killed worms also died. She noted the long-term persistence of chlorinated hydrocarbons in soil and the possible transfer of chemicals into plants grown in such soils. She stated that government officials had aerially sprayed areas without notifying the public, and that these officials underestimated the safety problems of chemicals. Carson highly praised the desirability and great potential of using biological controls in place of chemicals, as well as use of natural products and less toxic chemicals (e.g., pyrethrins). She pointed out that scientists' and government officials' concerns addressed only classical toxicity of pesticides and that no testing was done on effects to wildlife. Regarding residues in food, she stated that government protection through the Food and Drug Administration was minimal and that tolerances provided a false sense of security, because usually, only minimal safety data were available.

In human safety, Carson pointed out that exposure to or ingestion of various products, each at individually safe levels, taken together, could lead to health problems. Also, she described the concept of delayed physiological symptoms (e.g., mental problems and cancer). She also considered disruption of key metabolic pathways and mutations a high price to pay to have no mosquitoes. She stated that with safety knowledge increasing rapidly, what is safe today is not safe tomorrow. She cited tumors and leukemias brought on by carbamates, DDT, and aminotriazole as problems.

Carson discussed the resistance of insects to insecticides at length and indicated that the U.S. Department of Agriculture's solution at the time was to recommend more frequent sprays or greater quantities. She stated that DDT brought on the "age of resistance" and noted that chemical treatment was a treadmill that, once started, could not be stopped.

Carson concluded that our desire for total control of nature was conceived in arrogance.

In the Beginning

Rachel Carson in 1961 in Maine.

(Photograph by Erich Hartmann. Used by permission of the Rachel Carson Council, Inc.)

1 ∽ Rachel Carson: Her Vision and Her Legacy

Shirley A. Briggs

To assess Rachel Carson's accomplishments and their effects now, 20 years after her death, requires cutting through a good deal of curious mythology. Some has been constructed by her opponents, but much has

0980-4/87/0003$06.00/0 © 1987 American Chemical Society

come from her admirers as they tried to make Carson fit their preconceptions. We who knew her before and during her years of major achievement find these myths interfering with the understanding of what she actually did and why.

Rachel Carson's Motives

Carson was neither a mystic nor a recluse, and certainly not a reformer by nature. Those of us who followed her path have often been falsely classified as "antiestablishment, antitechnology, probably communist nuts". This epithet would not have applied to her either. She was a meticulous scientist and a writer whose highest purpose was to make the intricate and awesome totality of the natural world clear to all of her readers. She wrote with the same care for accuracy and precise wording that she applied to her search for the essential facts. Often commended for the beauty of her language, she felt that beauty was integral to science. When accepting the National Book Award in 1952, she said:

> The aim of science is to discover and illuminate truth. And that, I take it, is the aim of literature, whether biography or history or fiction; it seems to me, then, that there can be no separate literature of science.... If there is poetry in my book about the sea, it is not because I deliberately put it there, but because no one could write truthfully about the sea and leave out the poetry.

When Carson first perceived the problems posed by the new kinds and uses of pesticides, she tried to interest others in undertaking the book that would be *Silent Spring*. No suitable and willing person was found; indeed, it became clear that she should write the book. Her years in the U.S. Fish and Wildlife Service brought her into continual contact with the scientists who first noticed the unexpected effects of these postwar chemical products. The adverse consequences of older pesticides had a long history, but their effects had been relatively contained. The use of lead arsenate on orchards was far different from the aerial spraying of thousands of square miles of forest. Such large areas of wilderness had never been subjected to such exposure, now made possible because of new cheaper materials. Wildlife and fishery biologists first noticed the unanticipated effects of these pesticides on birds, fish, and other animals.

It was during those U.S. Fish and Wildlife Service years that John George, Clarence Cottam, and I first knew Carson. Cottam, then assistant director, set the standard for scientific excellence and integrity, with primary concern for the public good and no concessions to economic or political pressures. Carson was a part of this tradition, and these were the people who helped her gather material and cheered her on.

The adverse effects of pesticides were well known in environmental circles when Carson undertook the book. Her role was to put together a diffuse and variable body of data and to shape it into a clear account that would not only educate the public but also reach people in authority. Before *Silent Spring*, as George explained in 1972, people in positions of economic and political power had been

> ...able to dismiss ecologists as impractical visionaries, whose counsel was of no import. Miss Carson had sufficient impact to reach the highest levels of government, and these decision makers began to listen to the idea that ecological process was vital to life and to our own well-being. As a result, the ecologist was less easily dismissed as perhaps well meaning but completely unrealistic.

Carson's advantages in writing *Silent Spring* stemmed from her dual ability as scientist and writer and from the considerable reputation she had achieved in both fields through her writings about the sea. She was also by then financially independent, and from the ensuing record of harassment of some of her supporters who had vulnerable jobs to lose, we can see how important financial independence was.

Paul Brooks, Carson's biographer, summed up her motives and her method at a 10th anniversary celebration of the publication of *Silent Spring*:

> She did not herself say that "Here is a great opportunity to write a world-shaking book." She had other things she wanted to do. She had made a great reputation; she had one of the great best sellers of all time. But her own sense of responsibility made her feel that

she simply had to do something on this subject of pesticides...and after exploring in all sorts of directions she realized that it had to be done by her or by no one. She was not really...a crusader at heart. She said that one crusade in a lifetime is enough. But I find now among younger readers there are a great many who know her only from *Silent Spring*, who never heard of *The Sea Around Us*, and certainly never dreamt of hearing of *Under the Sea Wind*. I think that they miss something in understanding what she did. She did it because she was first of all a scientist and a writer, and she became a crusader on one particular occasion because she felt she had to.

...Instead of making a lot of broad generalities about what we are doing today on the environment, she had one subject that she knew well. She went not from the general to the particular, but from the particular to the general, and she started an ecological revolution which was not what she thought she was going to do. But she started it because she knew this one important thing so terribly well. The paradox is that the really hysterical reaction on the part of the chemical companies and so on was not really because "we can't sell so much DDT and our profits are going to fall." DDT is a very small part of all this operation. The really scary thing was that she was questioning the whole attitude of industrial society toward the natural world. This was heresy and this had to be suppressed. This is what caused a really hysterical reaction against *Silent Spring*, and it is the thing now that has made her almost a prophet among the young.

Reaction to *Silent Spring*

The mythology evokes a very different picture. To those who knew her, the claim that she was an unscientific, emotional polemicist was so off the mark that it was hard to see how anyone could read the book and come to such a conclusion. We had seen her slowly and painstakingly gather all possible evidence, check it meticulously, select only the most typical and substantiated examples, and ponder every word to be sure it was accurate. Many reacted without reading the book, of course, but others found its message so at odds with their convictions that they reacted emotionally and then concluded that the book must have been written with emotionalism. Others, having been told it was a wild-eyed, emotional book, managed to read it as such.

We wish that more people could have heard her speak about the book, so her calm voice could bely these impressions. In a speech given in December 1962, just after publication of *Silent Spring*, Carson countered:

Early in the summer, as soon as the first installment of the book appeared in *The New Yorker*, public reaction to *Silent Spring* was reflected first in a tidal wave of letters: letters to congressmen, letters to government agencies, to newspapers, and to the author. These letters continue to come, and I am sure that they represent the important and lasting reaction to the book. Then even before the book itself was published, editorials and columns in the scores and even hundreds had discussed it all over the country, reflecting, I am sure, the true importance of the subject. The early reaction in the chemical press was rather moderate, and in fact I have had, and still have, fine support from segments of the chemical press and some segments of the agricultural press. But in general, as was to be expected, the industry press was not happy, and by late summer the printing presses of the pesticide industry and their trade associations had begun to pour out the first of a growing stream of pamphlets and booklets designed to protect and repair the somewhat battered image of pesticides...

One obvious way to weaken a cause is to discredit the person who champions it. And so the masters of invective have been busy: I am a "bird lover", "a cat lover", "a fish lover", "a priestess of nature", a devotee of a mystical cult having to do with laws of the universe which my critics consider themselves immune to. Another well-known and much used device is to misrepresent my position and then to attack the things that I have never said. Now I do not want to belabor the obvious, because anyone who has really read the book knows that I do favor insect control in appropriate situations, that I do not advocate the complete abandonment of chemical control, that I criticize modern chemical control not because it controls harmful insects, but because it controls them badly and inefficiently, and because it creates many dangerous side effects in doing so. I criticize the present methods because they are based on a rather low level of scientific thinking. We really are capable of much greater sophistication in our solutions of this problem.

Reviewers with jobs at land-grant universities and like institutions were in a difficult position if they approved of *Silent*

Spring but knew that their colleagues disapproved. Therefore, many reviews included pat phrases, such as, "Of course she exaggerated and made mistakes, but in general she was on the right track." These face-saving phrases have taken on a kind of immortality; they turn up again and again from people who admit, when questioned, that they do not know of any inaccuracies or mistakes, but "so many people said there were some." We at Rachel Carson Council have yet to be shown a valid example, despite the sketchy state of much of the information available to her at that time. This accuracy shows the value of her very conservative approach, and indeed much present knowledge shows that she understated the potential hazards. She steered carefully for the reasonable middle ground, and while she was prepared for opposition from the pesticide industry, she was especially concerned lest some of the more fanatical antichemical cults would identify her with their extreme stands.

George summed up Carson's achievements in 1972: "We consider Miss Carson the greatest biologist since Darwin, and already the far-reaching effects of her work in our society and our value system are evident." Consider the similarities: Carson and Darwin both took a vast range of information and not only gave it form that provided a broad concept of the natural world, but also put it in terms that people could grasp, test against their own observations, and use as a base for their own view of mankind's place in nature. Although these new concepts upset some existing tenets and aroused controversy, they were essentially constructive and were directed toward greater understanding and harmony. They encompassed the past and the future, and meshed complexities into a structure with order and the flexibility to endure. *Silent Spring* has influenced many fields such as law, government, agriculture, economics, and education, and has changed people's perception of their world.

Carson is credited with making ecology a household word, but more importantly, she gave the public a sound and inspiring understanding of the natural order. To show how potent chemicals can affect the whole fabric of nature, she had to first clearly describe our web of life. This positive side of the book was of great concern to her. She did not want to write a recital of doom, but she wanted the reader to appreciate how vitally humans can damage their world. The lesson applies to far more than pesticides, as the impact of the book shows.

The Comprehensive View

Opposition to Carson's viewpoint goes beyond the economic interests her work seemed to threaten. The schism is more between those specialists with tunnel vision and those individuals who attend both to details and the whole picture. It is not easy to see beyond one's immediate concerns to the intricate processes of nature and society. *Silent Spring* showed that the general public, given the impetus, may have a greater ability to see the whole picture than do some highly trained experts.

The comprehensive view is crucial to the long-term solution of the problems Carson addressed. Too short a focus on the immediate situation all too often leads to the quick fix instead of the lasting, and most effective, solution. George has pointed out that very few people use insecticides primarily to kill insects. How many people have much use for a great heap of dead bugs? Rather, they are trying to improve their crops or garden, to make their property more attractive, or to remove a nuisance or even a possible health problem. Could the sudden need for an insecticide have been avoided by better information, better planning, or a bit of extra effort? Could the ultimate purpose have been realized by a different crop or better timing? Are less hazardous materials available? Have the effects on the whole community been considered? What are all the costs of a given procedure? Carson's purpose was to give us the information and the state of mind that would enable us to consider these questions wisely.

The ability of all kinds of people to grasp these broad principles and complex details impressed Carson very much. The Rachel Carson Council continues to receive the kind of inquiries Carson got, and we have been even more impressed by the ability of people in remote areas with little formal education to seek out detailed and technical literature and to master it. Carson showed these people that they could and must grapple with these matters if they are to protect themselves and function effectively. This understanding may be the most important legacy of *Silent Spring*.

A kind of elitism often comes into play when scientists write for the layman. The notion that the public can only understand "popularized", meaning oversimplified, science, still has adherents. The corollary is that the specialists are so much brighter than ordinary folk that only they can penetrate the mysteries of science. We question both that assumption and the idea that science does not need the support and comprehension of the public. Both are much needed.

Just as mythology obscures Carson's character and purpose, it clouds the current facts on pesticides and the role of the federal government. Many people still believe that all pesticides on the market have been thoroughly tested and certified as safe, but the National Academy of Sciences' (NAS) latest estimate (1) is that only 10% have been adequately tested to permit complete health hazard assessment. For 38% we have no toxicity information at all. A basic assumption encountered in proceedings to cancel a pesticide illustrates the short-focus approach that has led to so much pesticide misuse. Although we can find no such law, the Environmental Protection Agency (EPA) seems to believe that the federal government is required to certify a toxic chemical solution for every conceivable pest problem. EPA has been told that they cannot ban a dangerous and uncontrollable substance if there is nothing else registered for a particular use. But is this really necessary? Must a certain crop be grown under particular circumstances even if plants, soil, and water are to be seriously contaminated? Are there other ways to solve the problem, such as crop rotation or cultural methods? These alternatives never seem to occur to the special pleaders. Also, those who might profit from such substances are rarely the ones who would pay the costs of contamination.

Ironically, the EPA has by now so shifted its focus that testing of pesticides is done with such preoccupation for human exposure that it almost ignores broad environmental impacts.

A recent "Nova" program on National Public Television, "Down on the Farm", pointed to several reasons why so many farmers have difficulty with pesticides. The disciplinary fragmentation of science has caused disregard for the whole picture, and agricultural research has concentrated on short-term productivity heedless of the eventual results.

Solutions in the Carson tradition call on us to know more, pay closer attention, consider the future, and put in more personal effort. In return, we can often achieve more effective and lasting results without unwarranted side effects. There is an old-fashioned morality to this situation. Those who blindly seek the quick fix are often self-defeated.

In 1982, the Rachel Carson Council had a meeting to see how far we had come in the 20 years since the publication of *Silent Spring*. We looked at law, science, agriculture, economics, public understanding, and the scope of present pesticide use. For every advance toward sounder practices, we found serious setbacks. On the other hand, research continues into the third generation of more specific pesticides, and methods are being developed that require less broadly toxic chemicals and are more in accord with long-term human and environmental health. More of the public, often from hard experience, understand the need for better control of highly toxic products. As usual, the clear-cut, tangible issues, such as toxic waste dumps, get the most support and action. But the general pervasion of our habitat with little-understood poisons may pose a far more serious and intractable problem. This situation is the complex matter that Carson addressed. It persists despite her efforts; we hope that more people will continue in her tradition of understanding and education.

We still lack much of the information we need to act wisely. At the leisurely pace at which EPA's reregistration testing for older pesticides is proceeding, it will be many years before the NAS standards are met. The best we can hope for may be precise and sophisticated new products that could supersede our present troublemakers, and the application of integrated pest management that makes use of several alternate techniques.

In this possibility, I believe Carson would find her greatest hope today.

Literature Cited

1. *Toxicity Testing: Strategies to Determine Needs and Priorities*; National Academy Press: Washington, DC, 1984, pp 12, 117.

Regulatory Aspects

2 ∾ The Not So Silent Spring

John A. Moore

A citation attributed to E. B. White states

I am pessimistic about the human race because it is too ingenious for its own good. Our approach to nature is to beat it into submission. We would stand a better chance of survival if we accommodated ourselves to this planet and viewed it appreciatively instead of skeptically and dictatorially.

Beat nature into submission! *Silent Spring*'s first chapter, titled "A Fable for Tomorrow", starkly portrayed the futile consequences of such an approach to pesticide use. Accommodate ourselves to this planet! This is a recurrent theme, or plea if you will, that Rachel Carson presented in her book that has so strongly influenced many of our thoughts and actions. It may surprise you that Carson never advocated the cessation of chemical pesticide use. On page 12, she said:

It is not my contention that chemical insecticides must never be used. I do contend that we have put poisonous and biologically potent chemicals indiscriminately into the hands of persons largely or wholly ignorant of their potentials for harm.

She also stated, "I contend, furthermore, that we have allowed these chemicals to be used with little or no advance investigation of their effect on soil, water, wildlife, and man himself" (p. 13).

0980-4/87/0015$06.00/0 © 1987 American Chemical Society

Evolution of Government Regulation

The federal government first became involved in the pesticide arena in 1910 with the passage of the Insecticide Act, whose thrust was toward consumer protection. The consumer was the purchaser of the insecticide, and the protection was from mislabeled and misrepresented (fraudulent) products. For more than 35 years, in a market in which there were relatively few pesticides, this law seemed to serve the needs of the nation.

After World War II, however, agrochemical use rose dramatically as the nation addressed a broad range of concerns ranging from control of disease-bearing insects to increasing food production. The United States, flush with the victory of wars on opposite sides of the globe, confidently turned to the synthesis and manufacture of complex organic chemicals to win the war against pests. Spectacular success in vector control in the fight against malaria typified the good to be realized.

In 1947 Congress passed the forerunner of our current pesticide law: the Federal Insecticide, Fungicide, and Rodenticide Act (FIFRA). This landmark piece of legislation retained the principal thrust of protecting the consumer (the pesticide user) from fraudulent products (those that did not measure up to their promise of effective pest control).

In 1952 the potential for deleterious consequences to humans, who were served a food supply that had been augmented through the use of pesticides, was recognized in an amendment to the Federal Food, Drug, and Cosmetic Act. This amendment established procedures for setting tolerances (residue levels) for pesticides used in food, feed, and fiber.

A decade later *Silent Spring* was published and caught the attention of Carson's countrymen. For many people, *Silent Spring* was their introduction to ecology; for others, the book introduced the concept of the "safety" of chemicals as well as their effectiveness. People began to view their environment as something they shared with the rest of the members of the ecosystem. An awareness grew of the presence of life around us and with it came an awareness of bird kills, fish kills, and sickness associated with the use (misuse?) of certain pesticidal products. Although little was definitively known at the time, a nation began to ask itself some environmental questions and to appreciate the need for answers to these legitimate concerns. We consciously acknowledged our stewardship duties for the earth. The lack of scientific information about pesticides led to anxiety, if not paranoia, in some quarters about the unknown effects of these substances on human health and the environment. Without data, it was difficult to rationally discuss this increasingly emotional issue.

Congress responded by amending FIFRA in 1964. These amendments recognized both the valuable contribution that pesticides were making to the nation and the potential for deleterious effects from pesticidal use on invertebrates and plants that are important to humans. The amendments also directed that attention be given to safety considerations in the labeling of pesticides.

In some respects the high-water mark of the adolescence of the environmental movement was reached in 1970. In June of that year the nation engaged in Earth Day, a happening of unprecedented proportion for a cause that was as deeply felt as it was vaguely defined. The Environmental Protection Agency (EPA) came into being that same year and was given the authority to regulate pesticides and set tolerances.

In 1972, a decade after the publication of *Silent Spring,* Congress passed far-reaching amendments to FIFRA that specifically mandated the protection of public health and the environment as a guiding principle for the use of pesticides. The criterion for protection was framed as freedom from "unreasonable adverse effects", which are defined as "any unreasonable risk to man or the environment, taking into account the economic, social, and environmental costs and benefits of the use of any pesticide".

This amendment process made the pesticide statute a major public health law only 14 years ago. The statute speaks of accomplishing pesticide regulation through consideration of risks and benefits. The naively appealing notion of safety is set aside in recognition of the fact that absolute safety may not be a viable concept. In all subsequent legislative statements on FIFRA, Congress has retained the philosophy of accepting some risk in order to obtain significant benefits. EPA was asked to examine each case and take action whenever the risks outweighed the benefits. The final decision that evolves from any risk–benefit analysis is subjective. Wisdom is needed. As no single right answer exists, the decision is especially amenable to second-guessing.

Current Pesticide Registration Criteria

Responding to these dramatic amendments to FIFRA, EPA developed criteria for new pesticide registration that are clearly attuned to Carson's demands. EPA now reviews chemicals by examining experimental data on their effects on soil, water, wildlife, and humans. The recently issued *Pesticide Registration Requirements* clearly details these data needs. The requirements for the registration of a chemical for use on land crops grown for human consumption span five major categories: product chemistry, residue chemistry, environmental fate, wildlife effects, and toxicology. Data from two additional categories, reentry and spray-drift characteristics, are also commonly required.

Specifically, EPA's data needs in toxicology encompass acute toxicity data by three routes of exposure, eye and dermal irritancy, dermal sensitization, delayed neurotoxicity, repeat dose 90-day study in a rodent and nonrodent species, chronic feeding study (usually 1 year), oncogenicity studies in two species, teratogenicity studies in two species, two-generation reproduction study, gene mutation studies, structural chromosome effects, and general metabolism. In contrast, the 1962 practice called for acute toxicity by two or three routes plus eye and dermal irritation. If residues were detected on treated crops (detection limits of 1 in 100,000), then a chronic study may have been performed. Environmental effects were not addressed.

A controversial portion of pesticide registration involves those chemicals that were developed and registered prior to implementation of the extensive health and environmental requirements just described. A procedure evolved, somewhat painfully at times, that defines the criterion to be used and the process to be followed in evaluating these pesticides and their uses. The persistent, broad spectrum insecticides cited so negatively in *Silent Spring* such as DDT, Aldrin, dieldrin, heptachlor, and endrin, to name a few, have been banned or severely restricted in this country. In 1962 103 million pounds of organochlorine pesticides was produced; 35 million pounds was produced in 1983. However, approximately 600 other active ingredients currently remain in the "old-chemical" category. The pace of the evaluation process is of particular concern; however, the approach is clearly prioritized. EPA uses a ranking scheme that gives priority review to chemicals that are used on food crops in significant amounts; that is, chemicals for which there is likely to be extensive exposure. By 1985, this prioritization plan resulted in a completed review of those chemicals that represent 50% of pesticide poundage applied annually in the United States. There is no mood of complacency at EPA. Further acceleration of the old-chemical review process is in progress. Standards for 127 of these active ingredients have already been developed; 34 in the past 18 months.

During the past 14 years, several additional amendments have been made to FIFRA. In nearly all cases the changes have been oriented toward improving the quality of the scientific base upon which decisions are reached. The philosophy of risk–benefit evaluation has not changed.

The extensive requirements for current registration do not always provide clear insight as to appropriate action. For example, the technical capability to routinely analyze in parts per million, billion, trillion, or quadrillion clearly surpasses the toxicologists' and other scientists' ability to confidently interpret human or environmental risk. In these circumstances a regulator must turn to scientific judgment, science policy, and public policy to chart a consistent course that is open to scrutiny. The Office of Science and Technology Policy (OSTP) Cancer Guidelines, in whose preparation EPA was an active participant, help describe the data interpretation process; EPA risk assessment guidelines are being reviewed, revised, and expanded for a variety of end points, not just cancer.

The increased frequency of pesticides in ground water and surface water used for human consumption is an emerging issue. Restrictions in the use of certain pesticides based on climate, chemical properties, soil type, and water table are destined to become a more common occurrence. In some instances agricultural practices may also need reevaluation.

The magnitude of the task of defining and enforcing acceptable pesticide use has not diminished in recent years. Pesticide use has increased since the publication of *Silent Spring*. In fact, EPA estimates (1) that relative to 1962 the volume of pesticide ingredients used in the United States doubled by 1984 to a level of a little more than 1 billion pounds. Herbicides represent about two-thirds of total pesticide sales. Greater than 90% of all corn, cotton, soybean, and peanut acreage is treated with herbicides each year. Roughly one-quarter of the pesticides used are insecticides, and more than 80% of that use is directed at pests in corn, cotton, and soybeans. EPA economists project a plateau or at best modest growth for pesticide sales in the near future.

Alternatives to Pesticidal Chemicals

In the particularly captivating final chapter of *Silent Spring*, "The Other Road", Carson stated that we were (in 1962) at a place where two roads diverged and urged that we as a society choose the path "less traveled by", which embraced an extraordinary variety of alternatives. Each alternative involved the use of biological intervention in some way. Some of her suggestions can be compared with subsequent events:

- Broader use of sterile male insects. The early success in screwworm control was cited, and the possible use of this approach with fruit flies was mentioned. This approach was used successfully in eradicating a Mediterranean fruit fly outbreak in California a few years ago.

- The use of attractants and repellants. Carson also speculated on the chemical bases underlying this effect. Great strides have been made in our understanding of the effects of these materials on pest behavior. These naturally occurring substances have been used with varying degrees of effectiveness on a broad spectrum of insects, ranging from the pink bollworm on cotton to the common cockroach in the kitchen.

- Natural predators and microbial diseases. Carson cited the use of the milky spore (BP) as effective in control of the Japanese beetle. Another bacterial pesticide, *Bacillus thurin-*

giensis (BT), has become a significant factor in the control of certain moths.

- The selective use of attractants and chemical poisons. This suggestion is a tangible example of the acceptability, even by her strict standards, of chemical use provided one uses the minimal dose to effect the task.

The examples cited are real, but the pace of their development is still quite slow, and their market share is limited. However, one should not lose sight of the nature of many of the newer chemical pesticides being submitted for registration. The environmentally persistent, broad spectrum biocides characteristic of the 1950s and 1960s are being supplanted by selectively lethal chemicals. Current products are often developed from the basic knowledge of a biochemical process that is susceptible to chemical interference. This susceptible process is often unique to a class of pests rather than common to a broad array of plant or chemical forms. The forecast is for accelerated development of products based on increased knowledge of plant and insect physiology.

Perhaps most encouraging is the recent practice of developing a pest management plan in which chemical pesticides are only a part of a multifaceted scheme. The emergent success story of boll weevil control in cotton production throughout the Carolinas is most illustrative. Through the use of the chemical dimilin, which has selective larvicidal and chitin-inhibiting properties, early season spraying with conventional chemical insecticides is not needed. Natural predators of other cotton pests that used to be destroyed by these sprayings are once again successful in keeping these pest species in natural balance. Synthetic chemicals, pheromones in this case, are also used to objectively assess the quantity and location of pest species. Through these uses of chemical attractants or selective chemical pesticides, the broad spectrum insecticides have been reduced by 8 to 12 applications per year. This approach is clearly in concert with Carson's philosophy.

As the last two decades represented the heyday of the analytical chemist, the years leading to the birth of the 21st century may be the domain of biotechnology. Enhanced techniques realized through application of this knowledge will speed the general pace of product development. Most exciting is the prospect of crafting (grafting may be the better word), at the gene level, new or enhanced pesticidal characteristics on some of our best characterized microorganisms. These types of approaches are representative of the "other road" *Silent Spring* urged us to consider. Certainly keen insight and deliberate review of such products is a prerequisite to their use just as we continue such reviews of microbial products developed by conventional means. In these review processes we should use all of our scientific understanding and not allow selected bits of data to be extrapolated to forecast implausible omens that are superficially provocative, but fundamentally misleading, to the uninformed.

Conclusion

As stated earlier, progress has been made in investigating toxic properties as a condition of pesticide registration and, more recently, in reassessing older products. The general public needs to better understand that the specific circumstances of use dictate actual exposure, a parameter that is critically important in assessing human and environmental risk. Toxic properties in the absence of exposure do not indicate risk; in a similar vein, changes in patterns of use that diminish exposure may reduce risk to acceptable levels. Pesticide use is risk balancing. Because of Carson's book and subsequent events, society in general has become more aware of the potential for problems from pesticides. A result of that awareness campaign can be found on the container of any pesticide in the form of the pesticide label.

We often view the farm as the source of our problems with pesticides. "If farmers would only use common sense and follow proper procedures when using these substances," we say, "these problems wouldn't exist." However, too often the nonfarm community is guilty of using a double standard. We demand strict adherence to label directions from farmers and erroneously envision them generously broadcasting destruction across the land to get a marginal increase in crop yield. In fact, the farmer has an economic as well as a personal health and an environmental incentive to minimize the use of pesticides. On the other hand, we probably all know homeowners who are functional illiterates when it comes to reading pesticide labels and who still believe in the "rock 'em, sock 'em, knock 'em dead" school of pest control. This dualism in our attitude toward pesticide use needs greater scrutiny.

Silent Spring in many ways has acquired the status of a symbol in our struggle to protect public health and the environment from the dark side of our own indispensable ingenuity. The book is more often quoted than read. The title is more often invoked to make a point than its content examined to challenge our current attitudes and practices.

However, the impact of the book is undeniable as well as immeasurable. An entire nation, which was emerging from the apathy of the 1950s and just beginning to experiment with the heady activism of the 1960s, proved to be a fertile field for its message. What the book lacked in fact it made up for in intuition. By today's standards, it may have been short in scholarship, but it was and still is long on insight.

I believe *Silent Spring* is remembered today because it prompted us to ask the right questions. Its graphic presentation forced us to confront the issue: What is the proper role of pesticides in our society?

The quotation from E. B. White that introduced this chapter appeared on page ii of *Silent Spring*. On the same page appears the following contribution from Keats:

The sedge is wither'd from the lake,
And no birds sing.

These citations connote an omen that *Silent Spring* described with startling clarity. The message was heard loudly by some, a bit muted by others; nevertheless, it was heard by almost all. It was also heeded to the end that we have not yet realized "A Fable for Tomorrow".

There is no silence to *Silent Spring*.

Glossary

Active ingredient   The substance in a pesticide product designed to "hit" the target organism. Other ingredients in pesticide products are inert and serve ancillary purposes, for example, to dissolve the active ingredient to make it easier to apply.

Chitin-inhibiting properties   The capacity of a substance to interfere with the production of chitin, which is the horny material that forms part of the hard outer coat of insects and some other organisms.

Oncogenicity   The tendency for the development of tumors (malignant and benign) in organisms exposed to a chemical substance.

Pheromones   Chemical substances that are produced by animals and that stimulate a specific behavioral response in individuals of the same species, for example, sexual activity.

Teratogenicity   The tendency for the formation of birth defects in the offspring of pregnant organisms exposed to a chemical substance.

Vector   An organism (e.g., an insect) that acts as a carrier for an infectious disease.

Literature Cited

1. *Pesticide Industry Sales and Usage: 1984 Market Estimates*; Economic Analysis Branch, Benefits and Use Division, Office of Pesticide Programs, U.S. Environmental Protection Agency: Washington, DC, September 1985.

3 ᘒ The Science and Politics of Pesticides

C. F. Wilkinson

Our world is heavily dependent on chemicals. If we look around our homes, our offices, and our transportation vehicles; if we consider the quantity, variety, and quality of food available to us; if we consider the quality of our health and the drugs available to cure our ills; if we can see in total the thousands of chemicals that in some way touch our lives; then we will begin to appreciate the enormous impact chemicals have on each of us. Chemicals are an integral part of our lives. They have been irreversibly built into our technology and, by and large, have immeasurably improved the quality of our lives.

Although various chemicals were employed to control insects and other agricultural pests as early as the late 1800s, the real pesticidal potential of synthetic organic chemicals did not become apparent until about 1940 when DDT exploded onto the world scene. DDT was immediately hailed as a panacea for insect control. Its discoverer, Paul Muller, was awarded a Nobel prize, and the new "miracle" chemical was quickly accepted and used with enthusiasm. During the next few years the worldwide use of DDT resulted in enormous economic benefits through the effective control of a wide variety of agricultural pests and decreased human suffering through the control of mosquitoes and other insect vectors of human disease. Spurred on by the success of DDT, the chemical industry began an

0980-4/87/0025$06.50/0 © 1987 American Chemical Society

intensive search for other synthetic organic pesticides; and a steady stream of new insecticides, herbicides, fungicides, and other pesticidal products began to appear on the market. The search continues to this day, and the U.S. Environmental Protection Agency (EPA) currently estimates that approximately 600 basic pesticidal chemicals are marketed in some 45,000 to 50,000 formulations (1).

The use of pesticidal chemicals is unquestionably important to modern agriculture and many modern farming practices. Increased automation, new cultivation techniques, monocultures, and the development of new high-yielding crop varieties, which are the basis of the "green revolution", are made possible in part because of the availability of pesticides. Pesticides are synonymous with modern agriculture and provide the most effective and most economically efficient means of controlling the many thousands of species of insects, weeds, fungi, and nematodes that compete for our food. Without chemicals we could not continue to enjoy, at reasonable cost, the fruit, vegetables, and other agricultural commodities to which we are accustomed. The estimated 30% losses from pests that would occur in the absence of pesticides would spell economic and human disaster for many developing countries where millions of people face imminent starvation or exist on a suboptimal daily caloric intake. The latest figures for the world population—5 billion (2) and expected to rise to 8 billion in the next 40 years (3)—emphasize the continuing need to increase agricultural productivity worldwide.

The Changing Image of Pesticides

During the first 20 years of the pesticide revolution, the positive, beneficial aspects of chemical use were emphasized. As with any new technology, overuse and misuse occurred. Despite the first appearances of pest resistance to pesticides in the mid-1940s and recognition of some adverse effects on fish, birds, and other nontarget species, little serious thought was given to the potential long-term consequences of pesticides on human health and the environment.

Then, in 1962, Rachel Carson's *Silent Spring* appeared, and almost overnight the balance shifted. In her book, *The Apocalyp-*

tics, Edith Efron (4) identifies Carson as the first apocalyptic of national importance, the first of a number of noted biologists and ecologists like Paul Ehrlich, Barry Commoner, and George Wald who vociferously expressed the same view that planet Earth was a finite entity and was doomed unless we learned to control our technological abilities; some people thought that it was already too late.

In *Silent Spring,* Carson used pesticides symbolically to illustrate her view of the dire consequences of our continued efforts to master nature through technology. Despite numerous scientific inaccuracies and broad unsubstantiated conclusions, *Silent Spring* had an enormous impact on the way pesticides were viewed. For the first time, people were made aware of the hidden costs of pesticides and their potential for causing adverse effects on human health and the environment. The public felt betrayed, and science and technology, previously considered valuable allies, were seen as nature's enemies. Although we could still appreciate the benefits of pesticides in terms of crop production, we suddenly realized that we must begin to balance these benefits more carefully against the risks.

Pesticides thus became the initial focus of the U.S. environmental movement of the 1960s and, in retrospect, we all benefited from the increased public awareness that resulted. Since that time, however, a large segment of the public has become increasingly concerned with the potential risks of pesticides, and many scientists and representatives of the chemical industry now believe that the pendulum has swung too far. Pesticides continue to receive a measure of "bad press" that belies their contribution to the overall problem of environmental pollution and neglects due consideration of their substantial societal benefits.

For the last decade and a half an emotional belief, often verging on hysteria, has existed in the United States that society is being not so slowly poisoned by pesticides and other products of modern chemical technology. Despite the facts that we live longer and generally enjoy a much better quality of life than at any other time in our history, we have become a society consumed with fear and obsessed by the risks in our lives. We run the risk of becoming "a nation of healthy hypochondriacs," says scientist and author Lewis Thomas (5). Nourished by an eager, sensation-hungry press, a large segment of the public has indeed become fearful, angry, and confused; many people are highly suspicious of the profit-guided motives of the chemical industry and mistrustful of government efforts to protect it.

For almost two decades an indignant public has insisted on assurances of the "safety" of pesticidal chemicals and has demanded more stringent government regulation to protect people from the perceived threats associated with pesticides and other chemicals. Particularly in the case of pesticides, chemicals that are introduced deliberately into the environment, the public wants a regulatory policy that will prevent the release of materials that might pose a threat to health or the environment.

Faced with this emotional wave of antichemical, antitechnological sentiment, U.S. legislation directed toward the protection of human health and the environment from chemicals of all kinds has increased enormously since 1970. Approximately 30 such laws have been enacted including several major pieces of legislation such as the amendments to the Federal Insecticide, Fungicide, and Rodenticide Act (FIFRA); the Toxic Substances Control Act; the Safe Drinking Water Act; and the Clean Air Act. Most of this legislation has served a valuable purpose. Some, however, was enacted too hastily in response to perceived rather than actual needs and frequently was based on inadequate knowledge or incomplete scientific data. Regulations continue to become more complex and more difficult to interpret, and in some cases inconsistency among different agencies has done little to reassure the public that all is under control.

Few would argue the need for some form of restrictive legislation for pesticides. The major question, however, is how this goal can be achieved in a manner that encourages industrial research and development, that meets society's need for a continuing supply of high quality food, and that satisfies the public that government is meeting its responsibility to protect human health and the environment.

The Regulatory Decision-Making Process

When Carson wrote *Silent Spring*, the registration of a new pesticide in the United States depended mainly on proof of efficacy and demonstration of an acceptable acute toxicity to animals. Starting with the FIFRA amendment of 1972 and continuing to this day, registration requirements have become increasingly more stringent. In addition to detailed information on product chemistry, environmental fate, wildlife impacts, etc., registration requirements now include a thorough toxicological evaluation (often erroneously referred to as a "safety" evaluation) designed to detect potential adverse effects on humans.

Because, by definition, pesticidal chemicals are toxic to certain life forms and because all living organisms bear a striking resemblance to one another at a molecular level, many pesticides have the potential to cause adverse effects in humans and other nontarget species. Therefore, no pesticide is completely "safe"; a finite level of risk always exists. On the other hand, a pesticide can be considered "safe" in a practical sense if, under the conditions of its proposed use, its level of risk is judged negligibly (i.e., acceptably) low relative to the benefits that it provides.

At first sight, the procedure to decide if a given pesticide should be registered for use entails a relatively straightforward risk–benefit evaluation. Because benefits can usually be assessed, at least qualitatively, from the intended use of a pesticide, and because efficacy will ultimately be decided by its success on the market, the decision of whether to register a material depends heavily on an assessment of its risk potential.

Risk is a measure of the probability that an adverse effect will occur. For a pesticide, risk is a function of the intrinsic ability of the material to cause an adverse effect (e.g., acute toxicity, delayed neurotoxicity, or cancer) and the intensity and duration of exposure (dose); the dose is related to the circumstances under which exposure is expected to occur. Clearly, the risks to a pesticide applicator are considerably greater than those to a consumer exposed only to traces of pesticide residues in food. The regulator, therefore, needs to consider a series of questions such as who and how many people will be exposed, to how much will they be exposed, and by what route? Will the primary risk be through occupational exposure of factory workers, applicators, or farm workers; and if so are protective measures feasible? Or is exposure expected to involve the general public, specific ethnic groups, or regional populations as a result of residues in food or drinking water?

Information on the expected environmental effects is also considered. Can the pesticide possibly injure nontarget species such as fish or birds? Does it show unusual environmental stability? Is it likely to undergo biomagnification through the food chain or to leach into ground water? Even economic factors, such as possible impacts on foreign trade and the availability of alternative chemicals, are taken into consideration.

The final step in deciding whether to register a pesticide consists of carefully weighing the information on the risk potential of a given pesticide under the conditions of its proposed use against the expected benefits. This step is termed risk management. Although regulators often clothe their decisions in pseudoscientific terms, risk management is not a science. It involves a series of value judgments by the regulator that can be made only after careful consideration of all scientific and other factors.

Occasionally the potential toxicological risks associated with a given pesticide are of such magnitude that the pesticide should not be used under any circumstances. In this case, risk alone is sufficient to deny registration or to remove the material from the market. More often, however, a quantitative estimate of a chemical's overall risk potential is just one of a host of factors that must be considered in arriving at a regulatory decision.

Consequently, the process of reaching a regulatory decision on whether a given chemical should be registered as a pesticide and under what conditions it can be used involves a complex series of steps that carefully weigh the anticipated societal benefits of the chemical against its potential risks to human health and the environment.

Because the current U.S. regulatory process is dominated by considerations of potential risk and because present concerns are focused primarily on assessing the potential adverse effects of pesticides on human health (adverse environmental impacts are now generally considered of secondary importance), a discussion of our current toxicological assessment capabilities is appropriate.

Toxicology—A Science and an Art

The assessment of risk is the concern of the toxicologist. The commonly accepted definition of *toxicology*—the science that studies the adverse effects of chemicals on living organisms and assesses the probability of their occurrence—clearly establishes risk assessment and prediction as integral components of the discipline. Toxicology is partly a science and partly an art. The science of toxicology consists of observing the adverse effects of a chemical in a particular test species or in vitro model system and of gathering both qualitative and quantitative data to characterize the effect. The art of toxicology involves using an often limited scientific data base to predict the probability of the occurrence of a given adverse effect under a different set of exposure conditions (route, time, and dose) in a different species.

Not long ago toxicology was considered a branch of pharmacology and was restricted primarily to the medical profession. Its main concern was the diagnosis and treatment of human poisonings from drugs and natural products.

As a direct result of the claims of the environmental movement that we were poisoning ourselves and the biosphere with pesticides and other chemicals, toxicology began to expand in new directions. It assumed a new identity and became known more generally as environmental toxicology. For the first time, scientists, whose previous ideas of comparative studies were restricted to identifying differences between rats, mice, and occasionally guinea pigs, started to look more carefully at the effects of pesticides on fish, birds, and other wildlife and began to investigate the problems of biomagnification associated with many of the chlorinated hydrocarbon insecticides. Scientists began to wrestle with the problems of developing appropriate animal tests to evaluate the potential toxicological effects of pesticides on humans.

Toxicology has emerged as a bona fide multidisciplinary science that attracts the attention of chemists, biochemists, physiologists, geneticists, pathologists, and a host of other specialists who focus their combined expertise and research skills on improving our understanding of the interactions o. chemicals and living organisms. Our understanding of all aspects of the science of toxicology has increased immeasurably. Despite these advances, we still know little about how chemicals exert their toxic effects, and we have little or no capacity to predict the adverse effects of a given chemical in an intact animal, particularly in a human being. The art of toxicology is only as good as the science upon which it is based.

Toxicological Evaluation

Current toxicological testing requirements include a battery of acute and chronic tests on two or three animal species, usually rats, mice, and dogs. The tests provide data on acute oral and dermal toxicity, eye and skin irritation, carcinogenicity, teratogenicity, reproductive impairment, and neurotoxicity. Data from a variety of special tests, such as in vitro assays for mutagenicity, may also be required. Some of these tests can be completed in a relatively short time and require only a limited number of animals, but others, particularly the chronic oncogenicity and reproduction studies, require large numbers of animals and extend over several years or generations. These tests provide a continual record of the well-being of every animal throughout the test period and include a complete terminal pathology report on almost every tissue from nose to tail. The results should satisfy even the most skeptical person that every effort is being made to detect potential adverse health effects. But obtaining the toxicological data is only the beginning; the real question is how do we evaluate and interpret the data and directly relate the results to human risk?

Assessment of Acute Toxicity

According to World Health Organization and other estimates, up to 500,000 illnesses and as many as 20,000 deaths can be attributed annually to pesticidal chemicals worldwide (6). Acute toxicity, therefore, remains a matter of serious concern to pesticide formulators and applicators who, along with the victims of accident, misuse, and suicide, account for the majority of these casualties.

Although it does not receive the same degree of emphasis it once did, the assessment of acute toxicity resulting from single-dose exposures is still considered an important part of the toxicological evaluation process. From a practical standpoint, the determination of acute oral or dermal toxicity is a relatively simple task because it is hard not to conclude that at a certain dose level the animals either died or suffered some overt adverse toxicological effect. Unfortunately, the LD_{50} value (the dose that is lethal to 50% of test subjects), usually employed as a measure of acute toxicity, seldom helps in assessing risk because it provides no information on the slope of the dose–response curve and consequently is of little or no value in determining a toxicity threshold dose.

According to the fundamental tenet of toxicology, the severity of an adverse effect is related directly to the dose; therefore, a threshold dose below which no effect is likely to occur can be determined. In practice, the most important value that acute toxicity testing can provide is the lowest dose causing an observable effect; this value immediately focuses on the next lower dose that is defined as the "no observable effect level" (NOEL). The NOEL is the key to assessing acute toxicity because it is used to calculate the "acceptable daily intake" (ADI), defined as the amount of a material that can be ingested daily by humans over a lifetime and have no adverse effect. The ADI is obtained by dividing the NOEL from the most conservative subchronic animal test data by an arbitrary factor (often erroneously called a "safety factor") that may range anywhere from 10 to several thousand depending on the degree of uncertainty inherent in the data.

The magnitude of the uncertainty factor reflects the amount and types of data available, the number of species for which data are available, the nature of the toxic effect, whether or not the effect is reversible, etc. It is designed to take into account possible differences between animal and human responses as well as individual variations within the human population. If data are available on the effects of a given chemical on humans, the uncertainty factor will be relatively low. Because the uncertainty factor reflects the quality of the data base and state of knowledge at the time it was established, the uncertainty factor and the corresponding ADI may be modified appropriately as new data become available.

An established ADI for a given material provides regulatory agencies with a relatively firm toxicological bench mark on which to base risk assessment, tolerances, guidelines, and other policy decisions. Although scientific opinions sometimes diverge on the toxicological significance of a particular adverse effect upon which the NOEL is based (e.g., the depression of erythrocyte or plasma cholinesterase by carbamate or organophosphorus insecticides) or on the adequacy of the uncertainty factor used to arrive at the ADI, this general approach works quite well for materials causing only acute toxic effects.

Furthermore, the general public seldom seems greatly concerned about the acute toxicity of pesticides. The human psyche is apparently able to tolerate the possibility that, under certain circumstances, a given compound will lead to a rapid demise. What society will not tolerate is the possibility, no matter how remote, that long-term, low-level exposure to pesticides and other chemicals might ultimately lead to sinister chronic effects such as cancer, mutagenesis, or birth defects, which are generally considered the ultimate insults to human health. Public fears in this area have been greatly heightened in recent years by a vocal group of toxicological apocalyptics who claim that up to 90% of current human cancers can be attributed directly to pesticides and other synthetic chemicals (4).

Carcinogenic Risk Assessment

During the last few years, the assessment of carcinogenic risk has emerged as the very hub of modern toxicology; it presents innumerable problems and is beset by uncertainty and controversy at every step along the way. Uncertainty begins with our lack of understanding what causes cancer. We know that it occurs when, for some reason, the natural machinery for checking cell growth goes awry, when homeostatic mechanisms that control cellular balance break down. It is partly a natural disease, possibly related to endogenous imbalances associated with ionizing radiation, aging, or genetic makeup; but it can also be triggered by a multitude of exogenous factors including diet, life style, and exposure to naturally occurring and synthetic chemicals (7–9). As a result of the complex, uncertain, multifactorial etiology of cancer, some scientists question the value of conducting tests to evaluate the carcinogenic potential of single chemicals.

Disregarding this philosophical dispute, the theoretical and practical problems inherent in carcinogenic risk assessment are formidable. A quantitative assessment of cancer risk requires far more than a qualitative evaluation of carcinogenic potential. Unlike acute toxicity evaluation, in which the usual objective is to measure the severity of a specific adverse effect in individual animals, cancer risk assessment seeks to measure increases in the frequency of an event in a population. Furthermore, acute toxicity tests are concerned primarily with measuring overt adverse effects of relatively high doses of chemicals over short periods of time, whereas the ultimate objective of cancer risk assessment is to detect low probability events at low doses over long periods of time. Thus, the quantitative assessment of cancer risk relies not only on the science and predictive art of toxicology but also on the complex field of probability statistics.

Statistics show that, at a 99% level of confidence, a test on 10 animals might fail to detect a cancer actually affecting up to 37% of the test population; similar tests on 100 and 1000 animals might indicate no tumors even though they may actually occur at a frequency of 4.5% and 0.46% in the respective population. Viewed differently, a test involving 1000 animals can be expected to detect an effect at the 99% confidence level only if more than 5 animals are afflicted. The implications of this result are considerable because the introduction of a chemical that causes cancer at a rate of 5 in every 1000 human beings could lead to 1 million cases of cancer in the current population of the United States.

The well-known "megamouse" experiment designed to measure with precision (95% confidence) the effective dosage of 2-acetylaminofluorene producing a 1% tumor rate required approximately 24,000 mice (10). This experiment was an attempt to overcome the statistical problems inherent in detecting carcinogenic effects. In practice, of course, conducting routine tests with the massive number of animals required to detect cancers at such low frequencies with any degree of confidence is not feasible. Typical 2-year rat oncogenicity studies utilize 500–1000 animals divided into groups of 50 of each sex receiving either control (no dose) or treated (usually three or four dose levels) diets. Because of the statistical and practical limitations of measuring effects at low doses, the actual doses employed in oncogenicity testing are usually high, often at or approaching the "maximum tolerable dose". Although not measurable directly, the biological response to low doses can be extrapolated from the effects actually observed on high doses. Whether this high to low dose extrapolation is scientifically valid, and if so, how it can most appropriately be effected are topics of considerable scientific controversy and debate.

Although the true shape of the dose–response curve is not known at doses lower than those used in the experiment, it has been accepted for some time that carcinogens can exert an effect down to zero dose. In other words, it is assumed that no threshold exists below which an effect will not occur; consequently, all extrapolations of dose–response data must pass through zero. Despite continuing scientific dissent, this assumption is firmly embedded as fact in most, if not all, agencies charged with developing regulatory policy toward pesticides and other chemicals because of its conservative approach (11–13). The acceptance of this theory has challenged statisticians and scientists to derive models that can determine the high to low dose extrapolation.

During the last few years, several models, all with impressive names—one hit, multihit, multistage, probit, logit, and Weibull—have been developed, and debate has been considerable about which, if any, is most appropriate for the quantitative assessment of cancer risk (14). The availability of these models and the understandable demands of regulators for numbers upon which to base policy decisions have made the U.S. regulatory process increasingly reliant on the development of numerical estimates of human cancer risk. We are caught up in a numbers game and are accustomed to furious debate about cancer risks of 1 in a million, 3 in 10,000, etc.

The risk assessment process is further confounded by uncertainties in extrapolating cancer data from animals to humans. Pathologists continue to actively debate the nature of the cellular lesions that should be considered in the assessment (12, 13). Well-developed malignant tumors are relatively easy to identify, but how should the so-called benign tumors and the vast number of neoplastic nodules and foci be considered? Cancer is now recognized as a multistage process. Although many of these neoplastic lesions may undergo repair and not develop into tumors, they might be indicative of carcinogenic potential.

Yet another confounding factor in the risk assessment process is the high spontaneous incidence of certain tumors in specific tissues of some strains of test animals. How can it be ascertained whether a small increase in the incidence of such naturally occurring tumors is a true reflection of the carcinogenic potential of the test chemical or simply a nonspecific response to stress, diet, or the physical environment? Is the presence of a certain type of tumor in one particular tissue of a mouse more relevant to assessing carcinogenic potential in humans than a tumor occurring in some other tissue? More often than not the honest scientist has to answer "I don't know" to these and a host of other questions.

Science or Politics?

Many toxicologists are becoming increasingly concerned about the current obsession for developing numerical risk estimates from highly theoretical models based on so many controversial unsubstantiated assumptions. They are frustrated because statistical argument frequently overrules scientific logic and because the science and art of toxicology, which should constitute the foundation of cancer risk assessment, often play second fiddle to considerations of how well the model fits the data.

A qualitative assessment of carcinogenic potential is a scientific process; it becomes a political process as soon as it is described in numerical terms. This transformation occurs because the results of quantitative risk assessment are highly dependent on a number of policy decisions that determine not only the data used but also the way it is processed. Of particular concern to the toxicologist is the fact that quantitative risk assessment relies primarily on data obtained from only one animal test. Consequently, the assessment places undue emphasis on the results of one particular experiment and fails to give adequate consideration to what may be an extensive additional data base; this emphasis is contrary to the favored "total weight of evidence approach" that attempts to take into account all the available data.

Other policy decisions dictate how the experimental data are processed. Should the assessment be based on the incidence of one particular type of tumor in a single tissue or should the total number of tumors in one or several tissues be combined? Policy also determines the model to be used for the high to low dose extrapolation. Should it be the model that best fits the experimental data (i.e., that most scientifically valid), or should it be the one that provides the most conservative result (i.e., that most comfortable politically)? Our current risk assessment methodology is highly susceptible to the whims of political expediency. A cynic might be forgiven for maintaining that an appropriate risk number can be derived to support a preconceived regulatory position.

The report on saccharin published in 1978 by the National Academy of Sciences (15) provides an excellent example of the extent to which risk estimates can vary. The report concluded that, over the next 70 years, human bladder cancer in the United States resulting from a daily exposure to 120 mg of saccharin might range from 0.22 to 1,144,000 cases, a risk estimate spanning 8 orders of magnitude. More recently, risk estimates spanning about 5 orders of magnitude were obtained with respect to the carcinogenicity of ethylene dibromide (16). It is sobering to note that in this case the values were derived from a single toxicological data point rather than the total weight of evidence approach.

Although regulators are usually quick to point out that numerical estimates of carcinogenic risk are used only as guidelines in making policy decisions, the estimates are so variable that one must question their value even for this purpose. The numbers add a false sense of precision and certainty to what in reality are often very imprecise, very uncertain data. Cancer is a disease of the whole animal, and we should not allow numbers to cause us to lose sight of the biology of the problem.

Because many toxicologists firmly believe that practical thresholds do exist for carcinogens, carcinogenic risk should be assessed by the same NOEL-safety factor procedure used for chemicals causing acute toxic effects. This approach would certainly be more honest, and the results would probably be just as precise.

Scientific Uncertainty and Regulatory Conservatism

The quantitation of human cancer risk from the results of animal studies must be approached with great caution. Science cannot provide unequivocal answers to many important questions. Furthermore, it is unlikely that we will ever have completely satisfactory answers.

Uncertainty exists because the questions asked about chronic health effects typically exceed the capabilities of the science of toxicology; they go beyond the realm of biological and scientific certainty. As such they should be termed *transscientific* rather than scientific, according to Alvin Weinberg, past director of the Oak Ridge National Laboratory (*17, 18*). Toxicologists engaged in assessing chronic health risks of pesticides and other chemicals constantly come face to face with transscience and increasingly are placed in the uncomfortable position of having to answer questions within the uncertain framework it provides.

The fundamental problem facing toxicologists is that the chronic adverse health effects of pesticides and other chemicals cannot be verified by direct experimentation. Consequently, their assessment invariably requires the extrapolation of data obtained under one set of laboratory conditions to those likely to be encountered under another different set of conditions. We can observe and measure an increased incidence of liver tumors in a population of laboratory rats exposed to 500 parts per million of a given pesticide in its food for a lifetime, but how do we use this information to assess the risk of cancer in humans exposed intermittently to 0.01 parts per million of the same pesticide in their drinking water? In the absence of rigorous experimental proof, even the best scientists are stripped of their objectivity and are reduced to making quasirational intuitive judgments that frequently lead to disagreement and personal animosity. The public and the media are at a loss as to why the experts frequently disagree on toxicological issues and view such dissension with alarm and mistrust.

The situation is exacerbated by the often heated adversarial legal system that currently pervades the U.S. regulatory process. No simple "yes" or "no" answers typically exist to questions such as "Does it?" or "Doesn't it?", "Is it?" or "Isn't it?", and many outstanding scientists are often impotent in the face of a good lawyer. The flimsy arguments of transscience are easily destroyed in a court of law that demands simplistic answers.

Several important implications of scientific uncertainty strengthen society's concerns for the human health risks associated with pesticides but also tilt the regulatory process too far toward risk considerations.

In contrast to currently accepted principles of human justice under which a person is assumed innocent until proven guilty, a pesticide is generally considered hazardous until evidence to the contrary is available. However, it is never possible to establish that some adverse effect will not occur because scientific methodology is limited to the detection and measurement of observable phenomena. Absolute safety (the absence of any adverse effect) cannot be established experimentally, and a certain level of risk must always be associated with any chemical. Thus, in answer to the frequent question "Is it safe?" the toxicologist must always hedge even when no basis for concern exists.

In practice, the results of chronic toxicity bioassays are seldom clear cut, and conflicting data sets are the rule rather than the exception. Unfortunately, only one positive test is required to irrevocably brand a pesticide as a carcinogen, irrespective of whether nine other tests proved negative or the positive test was poorly conducted.

The usual regulatory attitude toward conflicting data is to accept the results of the positive test and either to ignore the negative data or to consider them inconclusive. Regulators have become so risk conscious that they sometimes are unwilling to accept the results of a negative test, so they continue to request additional confirmatory data. Because of the highly variable nature of chronic toxicity testing, it is entirely probable, even with the most innocuous chemical, that testing ad nauseum will sooner or later uncover some observation that causes concern. Thus, although prudence constitutes an appropriate regulatory attitude, overcautiousness can lead to the development of a highly conservative policy that is not always in the best interests of science or the public at large.

An unfortunate result of this attitude is that the pesticides considered to pose the greatest chronic health threats often are quite simply those for which the most toxicological data are available. The perceived health threats may have little relevance to true toxicological potential. We must be careful to avoid the development of a policy that tends to summarily remove these well-tested pesticides from the market and to replace them with alternative chemicals that are perceived to be less hazardous simply because fewer test data are available; no risks are associated with a pesticide for which no test data are available.

Although prudence and conservatism in the regulatory process are understandable, current U.S. regulatory philosophy often seems reminiscent of that of the knight in Lewis Carroll's *Through the Looking Glass*:

> "I was wondering what the mousetrap was for," said Alice. "It isn't very likely there would be any mice on the horse's back."
>
> "Not very likely, perhaps," said the knight, "but if they do come I don't choose to have them running all about. You see," he went on after a pause, "it's as well to be provided for everything; that's the reason the horse has all these anklets about his feet."
>
> "But what are they for?" Alice asked in a tone of great curiosity.
>
> "To guard against the bite of sharks," the knight replied.

Some people feel that we can never do too much in the way of toxicology testing and, like the knight, firmly believe that in developing regulatory policy we must, indeed, "be provided for everything." On the other hand, we do not have the unlimited financial and other resources that would allow us to conduct exhaustive toxicology testing on every chemical to which we are exposed. We must use our limited resources wisely and guard against the tendency to become overly concerned about the bites of imaginary sharks.

In recent years, cancer has become by far the most fearsome of these sharks, yet absolutely no evidence exists that we are in the midst of a human cancer epidemic. With the exception of lung cancer from cigarette smoking, most forms of human cancer have either declined or stayed more or less constant over the last 40 years (19). If we continue to focus our limited testing resources on the assessment of cancer risk, the danger that we will not give adequate attention to other adverse effects ultimately might prove to be a more serious threat to human health.

Compromise and common sense must prevail in our attempts to conduct toxicological evaluations of pesticides and other chemicals and to assess the potential human health risks associated with them. More basic research is needed to understand the mechanisms of toxic action because ultimately, basic research will prove the key to improving our predictive capabilities.

Public Perception and Acceptance of Risk

It has been said that the only safe airplane is one that never leaves the ground, preferably one that remains in a locked hangar on an unused airfield. In this sense the only safe pesticide is one that is never synthesized. Public acceptance of some measure of risk is an inherent component of the regulatory decision-making process toward pesticides. But the perception of risk and the level of risk that is acceptable is an extremely complex, highly subjective issue.

All of us take risks of one kind or another every day. We drive automobiles, travel in airplanes, climb mountains, smoke cigarettes, and expose ourselves to a multitude of "over-the-counter" drugs for headaches and other minor ailments. Some of these activities present substantial risks that can be estimated from actuarial data, yet they are accepted as an integral part of living. Why then are most individuals unable to accept in a similar manner a finite level of risk, often orders of magnitude smaller than the risks of some daily activities, from the traces of pesticides and other chemicals to which they are exposed?

One major reason probably lies in the fact that the public vastly overestimates our scientific abilities to predict adverse chronic health effects from pesticides and views the risk estimates generated during the policy-making process as precise figures. The public feels that, because the effects are known and predictable, they should be avoidable; that they are not avoided is typically considered inexcusable and unacceptable. Futhermore, everyday risks such as driving cars and crossing busy streets are voluntary risks, risks in which individuals have a choice and as a consequence enjoy some measure of personal control.

Many people have stopped smoking in recent years because they consider the risks too great. But in the case of pesticides in our food or water, individuals do not have the luxury of making a choice; the risks are involuntary and inescapable. Worse than that, unknown, unseen bureaucrats in regulatory agencies are making the choice for them by registering pesticides that, according to their calculations, pose a cancer risk of only 1 in 1 million. "Who are they to play God and to make the choice for me? What if I just happen to be the one? Why should I trust them?" Time and time again, the public is left feeling helpless and dissatisfied, caught between what it sees as an inherently evil corporate industry and an ineffective, politicized regulatory bureaucracy. Although perhaps a simplistic view of a complicated issue, the question of who makes the choice and for whom is undoubtedly a critical component of risk acceptance.

Because society's acceptance of some measure of risk is an important factor in the development of a sound, credible policy toward regulating pesticides, much greater effort must be made to more fully inform and educate the public in this area. Scientists and regulators must play a more active role in explaining the facts, particularly the uncertainties, associated with the science of toxicology and the reality of our risk assessment capabilities; they must clearly explain the true magnitude of the risks involved. Somehow we must redress the imbalance of the perspective that causes individuals who smoke two or three packs of cigarettes a day and who consume substantial levels of naturally occurring carcinogens in their normal daily diets to be truly fearful of the effects of trace residues of pesticides in pancake mixes or drinking water. And last, but certainly not least, we must continue to remind the public that pesticidal chemicals are largely responsible for the quantity, quality, and variety of the foods that, like so many other aspects of modern technology, are now taken for granted.

Chemicals of all types are an essential and integral part of modern society and are certain to increase in importance in the future. To learn how to use these chemicals to the maximum benefit of society and with minimum risk to human health and the environment is a challenge that dictates the unified efforts of industrial leaders, researchers, educators, government officials, and the public and will require the development of trust, understanding, and close collaboration among all concerned.

Glossary

Acute toxicity　Toxicity usually occurring shortly after exposure to a toxic agent (e.g., a few hours or days).

Biomagnification　The tendency of fat-soluble materials to be passed through food chains (webs) and to accumulate in the tissues of certain species.

Carbamate insecticides　Insecticides that are esters of *N*-substituted carbamic acid.

Chlorinated hydrocarbon insecticides　Insecticides containing chlorine, hydrogen, and carbon.

Cholinesterase　An enzyme normally present in erythrocytes (red blood cells), plasma, and other tissues. It is involved in the transmission of nerve impulses.

Chronic toxicity　Toxic effects (e.g., cancer) usually occurring weeks, months, or years after exposure to a toxic agent or as a result of long-term, low-level exposure.

Dose level　The amount of chemical administered to an animal, often described in terms of milligrams per kilogram of body weight.

Dose–response curve　The change in toxic response with changes in dose.

Homeostatic mechanisms　Internal (physiological) controls—checks and balances—that maintain the body in a fairly constant state.

Level of confidence　A statistical term that indicates the degree of certainty of data.

Maximum tolerable dose   The highest dose used in cancer testing.

Nematodes   Tiny worm-like organisms that inhabit soil.

Neoplastic nodules and foci   Initial pathological changes in tissue that might represent early stages of cancer.

Neurotoxicity   Toxic effects on the nervous system.

Oncogenicity   The tendency for the development of tumors (malignant and benign) in organisms exposed to a chemical substance.

Order of magnitude   A 10-fold difference; for example, 1 and 10 are an order of magnitude apart, 100 is an order of magnitude greater than 10, but two quantities are the same order of magnitude if one is no larger than 10 times the other.

Organophosphorus insecticides   Insecticides that are esters of phosphorus-containing acids.

Quantitative vs. qualitative analyses   Numerical vs. descriptive analyses.

Teratogenicity   The tendency for the formation of birth defects in the offspring of pregnant organisms exposed to a chemical substance.

Toxicity   The intrinsic quality of a chemical to produce an adverse effect.

Literature Cited

1. Johnson, E. *Environ. Prot. Agency J.* June 1984, 4.
2. Data from the World Population Institute reported in the *New York Times*, July 7, 1986.
3. Data from the Population Division of the United Nations reported in the *1986 World Almanac*; The Newspaper Enterprise Association: New York, 1985.
4. Efron, E. *The Apocalyptics: Politics, Science, and the Big Cancer Lie*; Simon and Schuster: New York, 1984.
5. Thomas, L. *New England J. Med.* **1975,** *293*, 1975.
6. Copplestone, J. F. In *Pesticide Management and Pesticide Resistance*; Watson, D. L. and Brown, A. W. A., Eds.; Academic: New York, 1977.

7. Ames, B. *Science* **1983,** *221,* 1256–1263.
8. Doll, R.; Peto, R. *The Causes of Cancer;* Oxford University: New York, 1981.
9. Higginson, J. *Science* **1979,** *205,* 1263–1366.
10. Society of Toxicology *Fundam. Appl. Toxicol.* **1981,** *1,* 28-128.
11. *Toxicity Testing: Strategies to Determine Needs and Priorities;* National Academy of Sciences: Washington, DC, 1984.
12. Office of Science and Technology Policy *Chemical Carcinogens: A Review of the Science and Its Associated Principles; Fed. Regist.* **1985,** *50,* 10371–10442.
13. National Toxicology Program *Report of the NTP Ad Hoc Panel on Chemical Carcinogenesis Testing and Evaluation;* U.S. Department of Health and Human Services: Washington, DC, 1984.
14. Krewski, D.; VanRyzin, J. In *Statistics and Related Topics;* Csorgo, M.; Dawson, D.; Rao, J. N. K.; Saleh, E.; Eds.; Elsevier/North Holland: Amsterdam, 1981.
15. *Saccharin: Technical Assessment of Risks and Benefits;* National Research Council, National Academy of Sciences: Washington, DC, 1978.
16. *Ethylene Dibromide,* Position Document 4, U.S. Environmental Protection Agency, Office of Pesticide Programs: Washington, DC 1983.
17. Weinberg, A. M. *Interdiscip. Sci. Rev.* **1977,** *21,* 337–342.
18. Weinberg, A. M. *Issues Sci. Technol.* **1985,** *II(1),* 59-72.
19. *1982 Cancer Facts and Figures;* American Cancer Society: New York, 1981.

Specific Environmental Effects

4 ∽ Assessing the Toxicity of Pesticides to Aquatic Organisms

D. R. Nimmo, D. L. Coppage, Q. H. Pickering, and D. J. Hansen

As aquatic biologists with interests in toxicology, we would abstract the message in *Silent Spring* as "While ridding the world of weeds, weevils, and webworms/Be careful of fins, fur, and feathers." Naturally, our primary personal interests are the "fins" and their ecosystem mates. Thus, we've paid close attention to Rachel Carson's thoughts on the impact of pesticides on aquatic resources.

Hindsight: *Silent Spring* and Aquatic Resources

In the United States, prior to the 1950s or 1960s, aquatic biologists believed that our streams, lakes, and coastal waters, because of their large volumes, could assimilate all the products of agriculture, manufacturing, and municipalities without significant effect. *Silent Spring* provided the catalyst that made us realize that this situation was not true and changed our attitude of relative indifference to one of real concern. Concern turned to public awareness, public pressure, and finally,

0980-4/87/0049$06.25/0 © 1987 American Chemical Society

legislation and programs that were designed to prevent further deterioration of our water resources.

The emphasis in *Silent Spring* was not on economics, such as the value of aquatic resources or the worth of terrestrial ecosystems. But in the two chapters devoted to water and aquatic communities, Carson did ascribe worth or value. In Chapter four she began: "Of all our natural resources, water has become most precious", and in Chapter nine she stated, "The inshore waters—the bays, the sounds, the river estuaries, the tidal marshes—form an ecological unit of the utmost importance."

Carson did not quantify the value of our aquatic resources; however, their worth can be estimated. Here are some figures and dollar values to consider. In the United States there are

- 84,656,000 acres of freshwater (1),

- 70,000,000 acres of wetlands (2),

- 148,000,000 acres of continental shelf and estuarine waters (3, 4),

- 53,900,000 persons participating in fishing (5), and

- 25,400,000 persons participating in shellfishing and shell collecting (5).

Also consider that, in the United States,

- $17.3 billion is spent per year on consumer goods related to recreational fishing activities (6),

- $7 billion is contributed to the Gross National Product per year from harvesting, processing, and marketing commercial fish stocks (7), and

- $2.2 billion is the annual income from commercial landings (8–10).

Clearly our aquatic resources are extensive and valuable. Carson raised not only our level of understanding of these resources but also raised our consciousness of the hazards of indiscriminate manufacture and use of persistent pesticides on these delicate ecosystems. This chapter describes the progression of attitudes and methodologies in the field of aquatic toxicology since *Silent Spring*. Topics will include some problems still encountered with pesticides as well as some advances in determining (and preventing) acute and chronic effects.

Heptachlor and a Host of Others: Still Found Where They Should Not Be

Because of Carson's warnings, the uses of many pesticides have been restricted or canceled. Although many of these actions were taken more than 10 years ago, some of these chemicals are still of concern.

In Chapter four, "Surface Water and Underground Seas", Carson wrote about the Rocky Mountain Arsenal near Denver. For years the facility was leased to a chemical company to manufacture pesticides for use by the military. She wrote the following: "About 1950, or eight years after the chemical company had begun operation, farmers began to report unexplained sickness among livestock; they complained of extensive crop damage; foliage turned yellow, plants failed to mature...."

In July 1984, D. R. Nimmo was requested to provide some advice and ideas for a work plan involving studies for mitigation of pesticide residues found in fish from lagoons located on the Rocky Mountain Arsenal grounds. Apparently, the fish living in the lagoons were trophy fish but could not be taken off the property because of the magnitude of their contamination.

White et al. (*11*) reported on elevated DDE and toxaphene concentrations in fishes and birds in the lower Rio Grande Valley, Texas. In 1981, endrin was discovered in ducks of the central flyway—the result of pesticide use on grasslands in Montana (*12*). Heptachlor, used to control the fire ant* (a species related to the Southern variety but a pest of pineapples), was found as a contaminant in milk in Hawaii (*13*). Of course, we still live with our 15-year-old nemesis, the polychlorinated biphenyls (PCBs), which are perhaps one class of

---

*The fire ant in Hawaii is the same genus as that in our Southern states. The species "farms" aphids, which in turn carry the pineapple wilt, a virus that kills the plants. The only way to control the problem is to control the ant.

chemicals that Carson had in mind as the "substances that nature never invented". These anecdotes serve to remind us that chlorinated hydrocarbon pesticides still reside in many parts of our environment.

Have Environmental Residues Declined?

Some pesticides are declining in environmental samples. The Southern California Coastal Water Research Project (14) showed that municipal and industrial wastewaters, river runoff, aerial fallout, and harbor discharges in 1976 contained only 5% of the DDT they contained in 1970. Also, PCBs entering via outfalls decreased by 90% from 1972 to 1975.

Residues are also decreasing in biota. As early as 1969–1970, Butler (15) noted declines of DDT and metabolites in marine mollusks from 15 coastal states. Schmitt et al. (16) reported that from 1969 to 1974 DDT and metabolites steadily declined in freshwater fish in U.S. waters; however, since 1974, the residues have remained relatively unchanged. The lesson learned from past mistakes is that we must be careful not to replace the DDT-type pesticides with new pesticides if the new are persistent, accumulate in tissues of animals, and are toxic long after completing their purpose.

History: Coming of Age of the "Safe Concentration" Concept

Silent Spring had no direct bearing on the direction of aquatic toxicology, but it definitely identified a need to increase research in this field, particularly in late 1960s and early 1970s. Because Carson raised concerns about the effects of pesticides on aquatic biota, the work of early scientists was greatly expanded. For instance, Ginsburg (17) and Sandholzer (18) wrote about the effects of DDT on fish and the Chesapeake Bay blue crab. Young and Nicholson (19) in 1951 first associated pesticides with problems of freshwater species. In 1953, Doudoroff et al. (20) wrote about the "Toxicity of Some Organic Insecticides to Fish". In the early 1960s, Butler and Springer (21) described the potential hazards to biota in a paper entitled "Pesticides—A New Factor in Coastal Environments". Butler and Springer recognized that pesticides, along with other abiotic factors, could influence aquatic communities. They noted that since the advent of modern-day organic pesticides in the 1940s, the nation's aquatic life had one more factor with which to cope.

A series of events commonly referred to as the "Mississippi Fish Kills" captured the attention of Carson. She wrote in Chapter nine, "In Louisiana, 30 or more instances of heavy fish mortality occurred in one year alone (1960) because of the use of endrin in the sugarcane fields." Technical reports about the effects of endrin, suspected as the causative agent, were published during the 1960s (22–24). These reports highlighted the infinitesimal amounts of endrin in water that caused adverse effects in fishes. Thus began the careers of a cadre of aquatic toxicologists first located at the Newton Fish Toxicology Station in Cincinnati, Ohio, and later at the U.S. Environmental Protection Agency (EPA) Laboratory in Duluth, Minnesota. Historically, *Silent Spring* ensured an expansion of pesticide research fron 1968 to 1970 in terms of personnel and equipment, just before the establishment of EPA.

Some additional contemporary authors identified other problems with pesticides. *Pesticides and the Living Landscape* by Rudd (25) gave many case histories about the ecological problems of pesticides. Also important was a paper by Woodwell et al. (26) on "DDT Residues in an East Coast Estuary: A Case of Biological Concentration of a Persistent Insecticide". This paper described one of the first classical cases of so-called food chain bioaccumulation. Finally, *Since Silent Spring* by Graham (27) addressed many related subjects, such as the history of the country's response to pesticides as problems and the lack of control of chemicals other than DDT.

Within 12–15 years after the publication of *Silent Spring,* (1) aquatic toxicology became recognized as a discipline of the sciences, (2) testing procedures were beginning to be standardized to ensure data quality, and (3) life-cycle toxicity tests were being used to predict "safe" concentrations. We will chronicle the last point.

As early as 1959, Tarzwell (28) suggested that longer-term tests were necessary so that short-term 96-hour tests could be used to predict safe pesticide concentrations under conditions of continuous exposure. Three years later, papers that were published showed that long-term testing was feasible and provided some notion of "standard" methodology. First Pickering et al. (29) reported that Delnav, an organophosphate known for its acute effects, also showed cumulative toxicity to fish if exposure lasted several weeks. Mount (23) reported the first "true chronic" test using bluntnose minnows tested with endrin for 291 days. In 1963, Allison et al. (30) reported a 2-year life-cycle test of DDT-treated rainbow trout.

Mount and Stephan (31) in 1967 proposed to establish acceptable toxicant limits (i.e., safe concentrations) based on a comparison of acute to life-cycle results for fish. Shortly thereafter, studies with pesticides showed that "application factors" computed by dividing the chronic limits by the acute values were in fact similar for specific pesticides for several fish species; therefore, application factors appeared useful for estimating long-term safe concentrations from the results of short-term acute tests on species not tested chronically (32).

The experiments that represented major breakthroughs in the field of aquatic toxicology during this period typically used pesticides as test materials. Although life-cycle testing with freshwater fishes was accomplished in the early 1960s (23, 30) the technology is still rather new for other species; a successful life-cycle test with a marine fish was reported only 8 years ago (33) and a life cycle was completed with a marine mysid (a shrimp-like crustacean) only 7 years ago (34). About three dozen diverse species are currently used in acute testing, but only about a dozen fish species have been used in life-cycle testing (see the list on page 55).

Happenings: Fish Kills and Environmental Laws

Have the concerns for pesticidal effects on fins, fur, and feathers been real or imaginary? According to an EPA summary of fish kills between 1961 and 1975 (35), pesticides ranked second, accounting for 18% of those kills whose causes are known. The cause of most (29%) fish kills is unknown, and 21% of all fish kills are caused by low dissolved oxygen (see Glossary); the remaining 32% of fish kills are caused by contamination from chemicals other than pesticides. Because

Species of Fishes Used for Early Life-Stage Toxicity Tests

Freshwater

Salmon, *Oncorhynchus* sp.
Trout, *Salmo* sp.
Char, *Salvelinus* sp.
Northern pike, *Esox lucius*
Fathead minnows, *Pimephales promelas*
Common white sucker, *Catastomus commersoni*
Channel catfish, *Ictalurus punctatus*
Bluegill, *Lepomis macrochirus*

Marine

Sheepshead minnow, *Cyprinodon variegatus*
Silverside, *Menidia* spp.
California grunion, *Leuresthes tenuis*
Gulf toadfish, *Opsanus beta*

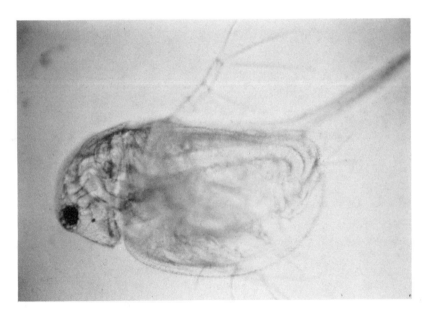

A freshwater species, the daphnid Daphnia pulex.

kills must be considered manifestations of acute effects, the chronic or long-term impacts of pesticides and other causes are largely unknown. Of the major fish kills (those involving more than 100,000 fish), insecticides as a category ranked as the sixth cause (3.7%) behind municipal wastewater (39.0%), unknown causes (18.1%), miscellaneous operations (12.2%), food processing (9.6%), and power plants (4.7%).

However, the effects of pesticides are perceived differently by different people. At a 1975 conference on marine pollution, two authors (36, 37) noted that in national surveys, concern about the effects of pesticides on the environment was ranked high by some, but not necessarily all, sectors of society. For example, Hann (36) reported that problems associated with pesticides in Texas were ranked only ninth out of 12 categories, preceded by such problems as pathogenic bacteria, organic chemicals, inorganic ions, and nutrients. Further, at the same conference in 1975, Li (37) showed that agricultural operations were responsible for only 7.8% of the fish kills in the nation's waters from 1963 to 1972. About 65% of these agricultural-related kills were due to pesticides.

An estuarine species, the pinfish Lagodon rhomboides.

An estuarine species, the spot Leiostomus xanthurus.

Evidence shows that the public's desire for strong environmental laws or standards for water has not diminished. (It could be argued that *Silent Spring* has kept the problems of pesticides and related toxic chemicals in the consciousness of Americans.) In a recent survey, 79% of persons interviewed were concerned with reducing water pollution, a value greater than those concerned with air pollution (*38*). According to the survey, 93% were against easing the Clean Water Act, 52% wanted it more strict, and 41% were satisfied with the present law. Only 4% wanted it less strict. Concerning toxics, 65% thought pollution of lakes and rivers was "very serious". The five most serious toxic problems, according to the results of the poll, were toxic pollution of rivers and lakes, leakage from hazardous waste, nontoxic pollutants, disposal of hazardous waste, and pollution of drinking water.

A 1984 poll showed (*39*) that rankings of the 12 most important environmental issues remained relatively unchanged from earlier results. The issues are listed by frequency (percentage):

- water pollution (26)
- air pollution (22)
- gun control (21)
- federal land sales (21)
- acid rain (20)
- antihunting (20)

- Indian rights (17)
- timber harvesting (14)
- mining (14)
- endangered species (14)
- Alaskan hunting land (14)
- grazing (13)

The continuous concern of the public has been responsible for legislation enacted to address water issues. The following federal programs have been enacted since 1970:

- Federal Insecticide, Fungicide, and Rodenticide Act (FIFRA)
- Water Pollution Control Act (Clean Water Act)
- Toxic Substances Control Act (TSCA)
- Marine Protection Research and Sanctuaries Act (MPRSA)
- Resource Conservation and Recovery Act (RCRA)
- National Environmental Policy Act (NEPA)

Although other environmental legislation has been enacted, these laws eventually resulted in many more regulations than procedures for testing aquatic species. FIFRA, which regulates the process of registration for use of a product, is listed first because it requires data from the most complete and complex sets of test methodologies and a variety of test species. The completeness of FIFRA was clearly an early response to the concern about pesticides. Some procedures required for conducting tests were adopted or modified for the other laws; however, the amount or complexity of tests is reduced as one proceeds down the list. Interestingly, of the initial National Criteria for water, only 21 contained sufficient data to derive numerical criteria, but 8 of the 21 were pesticides (40). Again, the amount of data necessary to derive criteria reflected past interest in pesticides.

How Little Is Too Much?

More than 20 years ago, Carson correctly predicted "And the early stages of these fishes (coastal water forms), even more than the adults, are especially susceptible to direct chemical poisoning." In the intervening years, investigators found that early life stages of freshwater and marine fishes indeed are more susceptible to most pollutants, including pesticides. McKim (41) reported on studies of 34 toxicants and four fish species in which the early life stages were compared to complete life-cycle chronic toxicity tests. Based on end points of hatchability, survival and growth, and increased deformities, in 82% of the tests the early life-stage estimates were identical to the life-cycle chronic estimates. In the remaining 18% of the test comparisons, the estimates for both deviated by a factor of 2 or less. In other words, as many fish died in early life stages as those that had completed their life cycles, so life-cycle tests may not be necessary in all instances. Furthermore, Ward and Parrish (42) reported that early life-stage tests of saltwater fish provided a close estimate of the "no effect" concentration of life-cycle tests.

In a test with a candidate pesticide on the early life stages of fathead minnows, the greatest effect was mortality between hatching and the 72 hours old; this effect continued throughout the 32-day test as the number of viable juveniles decreased with increasing concentration of the chemical (Table 1). In addition, the average lengths and weights of the juveniles were less with increasing concentration of the chemical (Table 2).

Some recent physiological and biochemical responses to many toxicants have been discovered. Perhaps Carson was alluding to this aspect when she wrote of "unseen, largely unknown, and unmeasurable effects of pesticides reaching our inland and coastal waters". Although the ecological implications of fish kills should concern us, the subtle nonlethal impact of

Table 1. Effects of Candidate Pesticide on Early Life Stages of Fathead Minnows

Pesticide Concentration[a]	Percentage of Fertile Eggs at 48 Hours	Percentage of Normal Larvae at 72 Hours	Percentage of Normal Juveniles at 32 Days
0.320	86.9	13.8	6.9
0.180	97.7	23.6	16.7
0.100	99.2	100.0	99.0
0.050	96.9	100.0	90.2
Control	95.4	100.0	100.0

[a]In milligrams per liter.

Table 2. Effects of Candidate Pesticide on Juvenile Stages
of Fathead Minnows at 32 Days

Pesticide Concentration[a]	Average Weight[b]	Average Length[c]
0.320	0.095	22.29
0.180	0.112	23.18
0.100	0.141	25.74
0.050	0.146	26.11
Control	0.151	26.13

[a]In milligrams per liter.
[b]In grams.
[c]In millimeters.

acute and chronic exposure to pesticides may result in ecological damage that is long lasting and rather severe but not easily detected. Effects on growth and reproduction of fishes, decreases in productivity of lower trophic levels in the food chain, and sublethal effects on physiological–biochemical functions could be part of this threat. For example, less collagen and hydroxyproline in the vertebral columns of fishes are early biochemical indicators of reduced growth and development due to toxaphene (43). Using these indicators of bone development (along with growth, reproduction, and survival), a 1977 study showed the acceptable limits for toxaphene were 25–54 µg/L for fathead minnows (*Pimephales promelas*) and 49–72 µg/L for channel catfish (*Ictalurus punctatus*) (44). A more recent study showed that hydroxyproline was significantly reduced in fish exposed to the organophosphate defoliant, DEF (45).

Reduced acetylcholinesterase (enzyme) levels in animal tissues have been a useful diagnostic tool for demonstrating pesticidal effects in laboratory and field studies on estuarine species. Sheepshead minnows (*Cyprinodon variegatus*) exposed to malathion for 26 weeks had reduced enzyme activity in their brains (46). Coppage and Duke (47) found the test useful in documenting exposure of several fish species to malathion following large-scale applications along the Gulf Coast. In another laboratory study, using fish in a partial life-cycle test, Goodman et al. (48) found that acetylcholinesterase activity varied inversely with the concentrations of the organophosphate insecticide Diazinon.

An estuarine species, the mysid Mysidopsis bahia.

Other important discoveries have been made in the past 20 years. For example, by extending the duration of the test or expanding its design to include end points such as those in early-life-stage studies, many significant end points can be revealed. For example, chlordecone (Kepone) was toxic to juvenile sheepshead minnows; the concentration that was lethal to 50% of the test subjects, referred to as the LC_{50}, was 69.5 μg/L after 96 hours of exposure (49). Couch et al. (50) described a pathological anomaly (scoliosis) in adult sheepshead minnows exposed to chlordecone at 4 μg/L for 10–17 days. Then, Hansen et al. (51) reported a 28-day LC_{50} of 1.3 μg/L for adult fish, and reductions in juvenile fish growth at concentrations as low as 0.08 μg/L. Finally, an entire life-cycle test of 141 days showed external signs of poisoning such as darkening of the posterior third of the body and decreased growth in fish exposed to as little as 0.07 μg/L (52).

Invertebrates as well as fishes can be affected in unusual ways by minute amounts of pesticides. For example, diflubenzuron (Dimilin), an inhibitor of the formation of an insect's rigid covering, was tested on an estuarine mysid (*Mysidopsis bahia*)

(53). The dose at which there was no effect was estimated to be between 0.015 and 0.025 μg/L. If the mysids matured in diflubenzuron at sublethal amounts, the second generation, though exposed only via their parents, produced fewer progeny and died sooner than unexposed controls. Furthermore, an abbreviated exposure to an acute concentration (24 hours at higher concentration) resulted in significantly fewer young compared to unexposed mysids. Other pesticides have been shown to be much more acutely toxic to this mysid—some had 96-hour LC_{50} values as low as 0.008 μg/L (8 parts per trillion) (54).

The generation of pesticides that concerned Carson has for the most part ended, although lindane and chlordane are still used. However, new chemicals with differing modes of action, different metabolic pathways, and increased potency have prompted new challenges and innovations for determining the ultimate impact on the aquatic environment. No longer will a couple of acute toxicity tests suffice; rather, a host of chemical, physical, and biological determinations including life-cycle testing must be made for complete risk assessment (55).

Henceforth

Pesticide use increased after *Silent Spring* and may well continue to increase in the future—certainly on a worldwide basis. One wonders how Carson would evaluate her contribution. How would she rate our present-day environmental problems? She would undoubtedly crusade about the dangers of toxic wastes and the contamination of ground water. She spoke of these problems more than 2 decades ago, but even she may not have seen the magnitude of the problem. About half of the U.S. population, including 95% of rural residents, depends on well water for domestic use (56, 57). Ground water consumption is increasing at twice the rate of the consumption of surface sources of water. In the semiarid West where options other than ground water for domestic use are limited, the situation is especially critical. Usually, the only alternative is to dam and divert free-flowing streams and use them as drinking water and thereby drastically change the historical use of the stream. If ground water contained, for example, one or two of the substances shown in Table 3, it would be less safe for aquatic life than for humans to use as drinking water.

Table 3. Comparison of Water-Quality Criteria for Several Pollutants

Pollutant	Safe Levels for Human Consumption	Safe Levels for Aquatic Life
Acrolein	320.0	21.0
Cadmium	10.0	6.3[a]
Cyanide	200.0	3.5
Endosulfan	74.0	0.056
Endrin	1.0	0.0023
Lead	50.0	20.0[a]
Mercury	0.144	0.00057
Nickel	13.4	160.0[a]
Pentachlorophenol	30.0[b]	3.2
Selenium	10.0	35.0
Silver	50.0	13.0[a]
Zinc	5000.0[b]	570.0[a]

NOTE: All values are given in micrograms per liter.

Conditions and estimates of exposure to these substances are different for humans and aquatic life; for example, for humans, estimates are based on an average intake of 2 liters of water per day; for aquatic life, estimates are based on a "no-effect" continuous exposure to the substances in a life-cycle test. An important consideration is that aquatic species are sensitive to concentrations of contaminants generally considered to be "safe" for long-term human consumption.

[a] Criterion established for a water hardness of 200 mg/L.
[b] Based on organoleptic (taste and odor) examination of tainted fish.
SOURCE: Reference 51.

Perhaps Carson would address the tremendous use-volume of herbicides in the United States; recent monitoring shows that herbicides are occurring in surface waters from agricultural areas (58). Although we do not have recent data on total use of pesticides, we suspect that the sevenfold increase in total pesticide use is largely due to the increase in this single category (herbicides). In the past, the tendency was to discount the effects of herbicides (e.g., Atrazine, Alachlor, Dual, 2,4-D, and Sencor) on fish and wildlife on the basis of their relatively low acute toxicities (i.e., high LC_{50} values), but what about the long-term effects on aquatic plants? The imperceptible reduction in the growth of emerging aquatic plants not only limits the production of energy necessary to sustain aquatic ecosystems, but also alters the physical habitat that provides "hiding places" for invertebrates and fish. Losses of aquatic plants greatly increase erosion, again, a significant alteration in the aquatic habitat.

Synthetic organic compounds will likely be used for some time against a variety of pests, but the use of environmentally persistent chemicals will probably diminish. Use of encapsulated slow-release formulations of relatively nonpersistent pesticides may ensure pest control for significantly longer periods, but this attribute may also facilitate chronic exposure to nontarget species. Additional chemicals or approaches will probably be more widely used, and this increased use will pose additional challenges for toxicologists to complete the hazard assessment process. Current testing procedures, apparently successful for first-, second-, and third-generation pesticides, may not suffice. Specific and unique modes of action of sterilants, pheromones, chitin inhibitors, and sporulating agents, as examples of third-generation agents, make many of the standard testing procedures inappropriate. Longer tests that last through multiple generations could be required; considerations of temperature, pH, photosensitivity of the test substance, methods of application, and metabolism will be as important as toxicological data. We will likely see an emphasis on the use of mesocosms or small replicated ponds or enclosures similar to those proposed by Boyle (59). These approaches are expected to help make more accurate determinations of the concentrations of pesticides that aquatic species will be exposed to under actual environmental conditions.

Finally, it could be that Carson would have taken up the issues of nuclear disarmament, the plight of the hungry in the Third World, or acid rain. If she had the chance, it is likely that she would have expanded the last chapter in *Silent Spring*, "The Other Road", which is in essence the outline behind integrated pest management (IPM) techniques. It is hoped that use patterns of multiple control agents and devices and use of a mosaic pattern of application instead of countywide coverage of a single chemical will be the rule instead of the exception. Perhaps we are finally beginning to understand and use ecological principles, and why shouldn't we mimic the best teacher around, Mother Nature? Carson said it so well:

> Through all these new, imaginative and creative approaches to the problem of sharing our earth with other creatures there runs a constant theme, the awareness that we are dealing with life—with living populations and all their pressures and counter-pressures, their surges and recessions. Only by taking account of such life forces and by cautiously seeking to guide them into channels favorable to ourselves can we hope to achieve a reasonable accommodation between the insect hordes and ourselves.

Glossary

Acute toxicity   A response to exposure that occurs in 48-96 hours.

Chitin inhibitors   Substances that interfere with the molting processes of insects (disrupt the formation of the chitinous "covering").

Chronic toxicity   A response to exposure that occurs over a long period of time, for example, several weeks, months, or even years.

Chlorinated hydrocarbon pesticides   Chemicals containing the elements of carbon, hydrogen, oxygen, and chlorine, for example, DDT and dieldrin.

First-generation pesticides   Chlorinated hydrocarbons.

Food-chain bioaccumulation   The concentration of pollutants from one trophic level to another, for example, from prey to predator.

Inorganic chemicals   Chemicals containing no carbon.

LC_{50}   Concentration of toxicant (e.g., pesticide) that is lethal to 50% of the test organisms in a specified time.

Low dissolved oxygen   Concentration of oxygen dissolved in water below which aquatic life cannot be sustained.

Metabolites   Compounds that differ chemically from the original pesticide and are formed as a result of organismic enzymes, sunlight, or other environmental factors such as pH and temperature.

Micrograms per liter (μg/L)   Parts per billion.

Milligrams per liter (mg/L)   Parts per million.

Organic chemicals   Chemicals containing the element carbon.

Persistent pesticides   Pesticides that remain in the environment and do not degrade or metabolize to innocuous constituents for months or perhaps years.

Pheromones   Naturally occurring substances (or their derivatives) that interfere with normal behavior of pest species.

Second-generation pesticides   Organophosphates (e.g., Malathion) and carbamates (e.g., Sevin).

Sporulating agents   Naturally occurring microbiological parasites that infect host pests. Control involves increasing infectivity of the spore within the host species. The best known example is Nosema, a spore used to infect grasshoppers.

Sterilants   Substances or treatments (e.g., radiation) that render insects sterile.

Third-generation pesticides   The newer types such as sterilants, pheromones, and chitin inhibitors.

Disclaimer

The views and opinions expressed in this paper are those of the authors and are not to be taken as the official policy of the U.S. Environmental Protection Agency.

Literature Cited

1. *Federal Aid in Fish and Wildlife Restoration Program;* U.S. Department of the Interior. EIS. U.S. Fish and Wildlife Service. U.S. Government Printing Office: Washington, DC, 1978.

2. Goodwin, R. H.; Niering, W. A. *Inland Wetlands of the United States;* U.S. Department of the Interior. U.S. Government Printing Office: Washington, DC, 1975.
3. Lauff, G. H. *Estuaries;* American Association for the Advancement of Science: Washington, DC, 1967.
4. Rounsefell, G. A. *Ecology, Utilization, and Management of Marine Fisheries;* C. V. Mosby: St. Louis, 1975.
5. *1975 National Survey of Hunting and Fishing and Wildlife—Associated Recreation;* U.S. Department of the Interior. U.S. Fish and Wildlife Service. U.S. Government Printing Office: Washington, DC, 1978.
6. *1980 National Survey of Hunting and Fishing and Wildlife—Associated Recreation;* U.S. Department of the Interior. U.S. Fish and Wildlife Service. U.S. Government Printing Office: Washington, DC, 1982.
7. *Current Issues in Natural Resource Policy;* Portney, P. R., Ed.; Resources for the Future: Washington, DC, 1982; 300 pp.
8. *Our Changing Fisheries;* U.S. Department of Commerce. National Oceanographic and Atmospheric Administration. U.S. Government Printing Office: Washington, DC, 1971.
9. *Economic Activity Associated with Marine Recreation;* U.S. Department of Commerce. U.S. Government Printing Office: Washington, DC, 1977; NOAA-S/T-77-2967.
10. *Fisheries of the United States, 1980;* U.S. Department of Commerce. National Oceanographic and Atmospheric Administration. U.S. Government Printing Office: Washington, DC, 1981; Current Fisheries Statistics No. 8100.
11. White, D. H.; Mitchell, C. A.; Kennedy, H. D.; Krynitsky, A. J.; Ribick, M. A. *Southwest. Naturalist* **1983**, *28*, 325–333.
12. Schladweiler, P.; Weigand, J. P. *Relationships of Endrin and Other Chlorinated Hydrocarbon Compounds to Wildlife in Montana, 1981–1982;* Montana Department of Fish, Wildlife, and Parks: 1983.
13. Smith, R. J. *Science* **1982**, *217*, 137–140.
14. *Annual Report for the Year Ending June 1976;* Southern California Coastal Water Research Project: El Segundo, CA, 1976.
15. Butler, P. A. *Pestic. Monit. J.* **1973**, *6*, 238–262.
16. Schmitt, C. J.; Ribick, M. A.; Ludke, J. L.; May, T. W. *National Pesticide Monitoring Program: Organochlorine Residues in Freshwater Fish, 1976–1979;* U.S. Department of the Interior. U.S. Fish and Wildlife Service. U.S. Government Printing Office: Washington, DC, 1983.
17. Ginsburg, J. M. *J. Econ. Entomol.* **1945**, *38*, 274–275.
18. Sandholzer, L. A. *Fish Mark. News* **1945**, *7*, 2–4.
19. Young, L. A.; Nicholson, H. P. *Prog. Fish Cult.* **1951**, *13*, 193–198.
20. Doudoroff, P.; Katz, M.; Tarzwell, C. M. *Sewage Ind. Wastes* **1953**, *25*, 840–844.
21. Butler, P. A.; Springer, P. F. *Trans. 28th North Amer. Wildl. Nat. Res. Conf.* **1963**, 378–390.
22. *Report on Investigation of Fish Kills in Lower Mississippi River,*

Atchafalaya River, and Gulf of Mexico; U.S. Department of Health and Welfare (Public Health Service). U.S. Government Printing Office: Washington, DC, 1964; p 22.

23. Mount, D. I. *Chronic Effects of Endrin on Bluntnose Minnows and Guppies;* U.S. Department of the Interior. (Fish. Wildl. Serv. and Bur. Sport Fish. Wildl.) Research Report 58, 1962.

24. Mount, D. I.; Putnicki, G. J. *Trans. Am. Wildl. Nat. Res. Conf.* **1963,** 177–184.

25. Rudd, R. L. *Pesticides and the Living Landscape;* University of Wisconsin Press: Madison, 1964; p 320.

26. Woodwell, G. M.; Wurster, C. F., Jr.; Isaacson, P. A. *Science* **1967,** *156,* 822–823.

27. Graham, F., Jr. *Since Silent Spring;* Houghton Mifflin: Boston, 1970; p 332.

28. Tarzwell, C. M. *Trans. N. Am. Wildl. Conf.* **1959,** 132–142.

29. Pickering, Q. H.; Henderson, C.; Lemke, A. E. *Trans. Am. Fish. Soc.* **1962,** *91,* 175–184.

30. Allison, D.; Hallman, B. J.; Cope, O. B.; VanValin, C. C. *Science,* **1963,** *142,* 958–961.

31. Mount, D. I.; Stephan, C. E. *Trans. Am. Fish. Soc.* **1967,** *96,* 185–193.

32. Eaton, J. G. *Water Res.* **1970,** *4,* 673–684.

33. Hansen, D. J.; Schimmel, S. S.; Forester, J. *J. Toxicol. Environ. Health* **1977,** *3,* 721–733.

34. Nimmo, D. R.; Bahner, L. H.; Rigby, R. A.; Sheppard, J. M.; Wilson, A. J., Jr. In *Aquatic Toxicology and Hazard Evaluation,* ASTM STP 634; Mayer, F. L.; Hamelink, J. L., Eds.; American Society for Testing and Materials: Philadelphia, 1977; pp 109–116.

35. *Fish Kills Caused by Pollution* (15-year summary 1961–1975); U.S. Environmental Protection Agency. U.S. Government Printing Office: Washington, DC, 1978; EPA-440/4-78-011.

36. Hann, R. W. In *Estuarine Pollution Control and Assessment;* Office of Water Planning and Standards. U.S. Environmental Protection Agency. U.S. Government Printing Office: Washington, DC, 1975; pp 310–329.

37. Li, M. In *Estuarine Pollution Control and Assessment 2;* Office of Water Planning and Standards. U.S. Environmental Protection Agency. U.S. Government Printing Office: Washington, DC, 1975; pp 451–466.

38. *Inside E.P.A. Weekly Report,* **1982,** *3,* 9.

39. Conley, C. *Outdoor Life* **1984,** *Nov.,* 1.

40. *Fed. Regist.* **1980,** *45,* 79318–79378.

41. McKim, J. M. *J. Fish Res. Bd. Can.* **1977,** *34,* 1148–1154.

42. Ward, G. S.; Parrish, P. R. In *Aquatic Toxicology,* ASTM STP 707; Eaton, J. G.; Parish, P. R.; Hendricks, A. C., Eds.; American Society for Testing and Materials: Philadelphia, 1980, pp 243–247.

43. Mayer, F. L., Jr.; Mehrle, P. M.; Dwyer, W. P. *Toxaphene: Effects on Reproduction, Growth, and Mortality of Brook Trout;* U.S. Environmen-

tal Protection Agency. U.S. Government Printing Office: Washington, DC, 1975; EPA-600/3-75-013.
44. Mayer, F. L., Jr.; Mehrle, P. M.; Dwyer, W. P. *Chronic Toxicity to Fathead Minnows and Channel Catfish;* U.S. Environmental Protection Agency. U.S. Government Printing Office: Washington, DC, 1977; EPA-600/3-77-069.
45. Cleveland, L.; Hamilton, S. J. *Aquat. Toxicol.* **1983,** *4,* 341–355.
46. Holland, H. T.; Lowe, J. I. *Mosq. News* **1966,** *26,* 383–385.
47. Coppage, D. L.; Duke, T. W. In *Proc. 2nd Gulf Coast Conf. Mosq. Suppression Wildl. Man.,* New Orleans, 1971, pp 24–31.
48. Goodman, L. R.; Hansen, D. J.; Coppage, D. L.; Moore, J. C.; Matthews, E. *Trans. Am. Fish. Soc.* **1979,** *108,* 479–488.
49. Schimmel, S. C.; Wilson, A. J., Jr. *Chesapeake Sci.* **1977,** *18,* 224–227.
50. Couch, J. A.; Winstead, J. T.; Goodman, L. R. *Science,* **1977,** *197,* 585–587.
51. Hansen, D. J.; Goodman, L. R.; Wilson, A. J., Jr. *Chesapeake Sci.* **1977,** *18,* 227–232.
52. Goodman, L. R.; Hansen, D. J.; Manning, C. S.; Faas, L. F. *Arch. Environ. Contam. Toxicol.* **1982,** *11,* 335–342.
53. Nimmo, D. R.; Hamaker, T. L.; Moore, J. C.; Wood, R. A. In *Aquatic Toxicology,* ASTM STP 707; Eaton, J. G.; Parrish, P. R.; Hendricks, A. C., Eds.; American Society for Testing and Materials: Philadelphia, 1980, pp 366–376.
54. Schimmel, S. C.; Garnas, R. L.; Patrick, J. M., Jr.; Moore, J. C. *J. Agric. Food Chem.* **1983,** *31,* 104–113.
55. *Estimating the Hazard of Chemical Substances to Aquatic Life;* Cairns, J., Jr.; Dickson, K. L.; Maki, A. W., Eds.; ASTM STP 657; American Society for Testing and Materials: Philadelphia, 1978; p 278.
56. *Water Well J.* **1981,** *35,* 58–59.
57. *Pesticides: EPA's Formidable Task to Assess and Regulate Their Risks;* U.S. General Accounting Office Report to Congressional Requesters; GAO/RCED-86-125; U.S. Government Printing Office: Washington, DC, 1986, p 138.
58. Butler, M. K.; Arruda, J. A. In *Perspectives on Nonpoint Source Pollution;* U.S. Environmental Protection Agency, Office of Water Regulation and Standards; U.S. Government Printing Office: Washington, DC, 1985, EPA 440/5-85-001.
59. Boyle, T. P. In *Aquatic Toxicology and Hazard Assessment: Sixth Symposium,* ASTM STP 802; Bishop, W. E.; Cardwell, R. D.; Heidolph, B. B., Eds.; American Society for Testing and Materials: Philadelphia, 1983; pp 406–413.

5 ∽ Impact of Pesticides on Ground Water Contamination

Robert F. Carsel and Charles N. Smith

M ore than 20 years ago Rachel Carson focused public attention on the potential effects of pesticides on the environment. Her book, *Silent Spring*, dramatically portrayed the results of environmental contamination by pesticides on the global ecosystem. *Silent Spring* brought the use of pesticides and their effects on surface waters and populations of nontarget organisms into public view. Even the term "ground water" was virtually unknown to the general public when *Silent Spring* was written. Carson also identified the potential interaction of pesticides and ground water contamination (Chapter 4, page 51): "...Soil becomes poisoned as a result of heavy application of arsenical pesticides.... Rains then carry part of the arsenic into streams, rivers, reservoirs, as well as into the vast subterranean seas of ground water."

In addition to farm-generated pesticides, Carson also implicated chemicals from hazardous waste sites as ground water contaminants (Chapter 4, pages 40, 42, and 44):

> ...A variety of pollutants combine to produce deposits referred to as "gunk".... Holding ponds become chemical laboratories for the production of new chemicals.... Ground water becomes contami-

0980-4/87/0071$06.00/0 © 1987 American Chemical Society

nated from this dump with the wastes traveling underground to nearby wells.... There is nothing probably more disturbing than the threat of widespread contamination of ground water.

Ground water contamination by pesticides, however, did not become an environmental issue until almost 2 decades after the publication of *Silent Spring* when widespread aquifer contamination by the pesticide aldicarb was found in the potato-growing region of Long Island, New York (1). This incident of ground water contamination was particularly significant because most area residents obtained their drinking water from privately owned shallow wells that tapped the aquifer.

An earlier example of contamination of ground water was reported in 1962 (2); 225 wells were found to contain compounds similar to phenoxy-type herbicides. These wells were thought to have been contaminated as a result of leakage from a nearby waste disposal basin.

These examples of drinking water contamination are among the reasons that ground water protection is one of today's major environmental issues. The need for ground water protection is compounded by the steady increase in the use of ground water for human consumption and for irrigation of farmland in semiarid and arid regions. In fact, withdrawal of ground water has tripled since 1950; the resource accounts for 25% of all fresh water used in the United States (3).

Current pesticide application technologies, land management practices, and pesticide properties also increase environmental concerns. New technologies that mix pesticides in irrigation water (chemigation) are used to apply agricultural chemicals to various crops. Conservation tillage, which leaves a protective crop residue on the soil, is used to reduce runoff and erosion, but it promotes the infiltration of pesticides into the soil profile. Soluble pesticides that partition (mix) into the water phase of the soil profile have replaced relatively insoluble chemicals.

The potential for aquifer contamination by leakage of chemicals, either disposed of or stored in landfills, lagoons, and holding ponds, is an additional environmental concern. Many such sites are found throughout the United States.

Ground water contamination, therefore, is among the most

complex issues currently facing environmental scientists and regulators. Because of its location beneath the water table, ground water presents especially difficult problems in detecting and monitoring contaminant levels and in evaluating the fate and effects of the pollutants. Remedial actions are difficult and expensive.

This chapter provides an overview of pesticide use and ground water contamination since the publication of *Silent Spring*, describes the progress in evaluation of this environmental problem, and projects some potential future issues that will have to be addressed.

Pesticide Leaching from Agricultural Fields

The vast majority of pesticides produced and used for agriculture when *Silent Spring* was published have either been discontinued because of efficacy problems or environmental concerns or restricted to specific nonagricultural uses. Any current direct investigation of ground water contaminated by these older discontinued pesticides that involved the monitoring of wells in agricultural areas would be difficult to interpret. Another means to evaluate past and present pesticide residues would be to conduct a literature search of reported ground water contamination. One such literature review, however, did not reveal appreciable contamination of ground water by pesticides produced in the past (2), partly because research emphasis has been on surface waters and nontarget organisms.

A solution would be to compare the relative potential for ground water contamination by pesticides from the past and pesticides currently used by developing a comparative methodology from known physicochemical properties. The objective would be to rank pesticides from past to present using a common reference that would indicate either a low or a high potential for ground water contamination based solely on the pesticide's mobility. (This approach would yield a conservative qualitative evaluation because transformation or decay would be ignored.) The methodology could be used in combination with current literature on pesticides in ground water to further evaluate the problem.

Description of the Problem

Pesticides can enter ground water from both nonpoint sources (e.g., an agricultural watershed) and point sources (e.g., a hazardous waste site). Nonpoint sources are characterized by highly variable loadings; rainfall dominates the timing and magnitude of pesticide infiltration into the ground water. Point sources are much less varied; the loadings are thought to be steady inputs into the ground water.

The potentially widespread nature of nonpoint source contamination makes remedial action difficult because it does not result in single streams that can be isolated and controlled, as can those that emanate from a point source. Prevention or reduction of future contamination must be based on understanding the relationships among chemical properties, soil system properties, and the climatic and agronomic variables that can combine to induce or inhibit pesticide leaching.

Many investigations (4–8) have shown that a pesticide's solubility in water, sorptive properties, chemical formulation, and soil persistence largely determine its leaching potential. Similarly, the important environmental and agronomic factors that determine a pesticide's leaching potential include soil properties, climatic conditions, crop type, and cropping practices. In short, elements of the hydrologic cycle interact with the chemical properties and characteristics of pesticides to transform and transport them within the unsaturated zone and into the ground water.

The most important characteristic of the hydrologic cycle that influences the migration of pesticides within the unsaturated zone is *recharge* (the flux of water that enters the water table). Recharge can be estimated for a unit of cross-sectional area of the land's surface by the *water balance equation*:

$$L + P = E + R_s + S + R \tag{1}$$

where L is the hydraulic loading of man-made sources (e.g., irrigation), P is precipitation, E is evapotranspiration (loss of water from the soil by evaporation and by transpiration from growing plants), R_s is surface runoff, S is the change in storage, and R is recharge. All values are in centimeters per year.

The driving force that largely determines recharge is total precipitation, which varies considerably from region to region. Man-made inputs such as irrigation are generally confined to the West, where evapotranspiration demands normally exceed rainfall, and to parts of the Southeast. The influence of irrigation on total recharge is generally small because most irrigation is designed only to meet evapotranspiration demands; however, some practices such as flood irrigation and spraying to prevent frost damage can contribute significantly to local recharge. The storage capacity of the unsaturated zone generally remains constant, particularly when averaged over a year or longer. Therefore, its change can be effectively dropped from the equation.

Because most pesticides are applied on or just beneath the soil surface, the rainfall–infiltration–runoff process must be quantitatively described to evaluate recharge in the context of potential pesticide leaching to ground water. To accomplish this evaluation, the water balance equation is simplified (unless specific irrigation practices dictate differently) as follows

$$R + P - R_s - E \qquad (2)$$

The total annual recharge estimated from Equation 2 can vary from less than 10.0 to more than 90.0 cm/year for different regions of the United States and is different from year to year.

Interaction with this recharging water largely determines a pesticide's transport and fate characteristics. The sorptive properties of pesticides significantly affect their physical movement because they determine the distribution of the chemical between the solid and aqueous phases of the soil matrix. Pesticides distributed in the aqueous phase can be carried in the recharging water to ground water. The sorptive properties of nonpolar pesticides generally correlate well with the organic carbon content of soil.

The available water-holding capacity of soils at any particular time is affected by such properties as temperature, humidity, soil texture and structure, organic carbon content, and plant–crop characteristics. The amount of water that exceeds this available storage capacity (after correction for runoff and evapotranspiration) will infiltrate down through the soil profile and eventually reach the ground water (unless it is intercepted by tile drains or impermeable layers above the water table). The *field capacity* of a soil, defined as the moisture content after gravity drainage has ceased, is less for sands and sandy loams than for clays and clay loams. The ability of pesticides to migrate with recharging water is greatest for soils with low water-holding potential. The *bulk density* of soils generally ranges from 1.0 to 2.0 g/cm^3; 1.25 to 1.75 g/cm^3 is common to sands and sandy loam soils (9).

The sorptive properties of pesticides and the water-holding and density characteristics of soils can be used to rank the relative mobility of pesticides with respect to water by estimating a *retardation factor*. The retardation factor provides a relative indicator of mobility and hence ground water contamination. Water has a retardation factor of 1.0. A pesticide with a retardation factor of 2.0 would possess mobility one-half that of water. The retardation factors of older pesticides are very high, so their leaching potential is very low. However, the retardation factors of older pesticides have decreased with time. For present-day pesticides, retardation factors can approach 1.0, so their leaching potential is high (i.e., the pesticides would move with the recharging water front through the soil).

Since *Silent Spring*'s publication, the leaching potential of most commonly used pesticides has increased. Most of this increase has occurred since 1970 with the advent of carbamate pesticides, which are very water soluble.

Specific Cases

A 1974 review of ground water contamination from pesticides and other chemicals reported the contamination of a shallow-bored well by DDT and toxaphene (2). It was discovered that pesticide-contaminated soil had been used as a backfill, and thus the contamination was caused by the movement of contaminated soil to the ground water instead of by the usual contaminated recharge-water pathway. In the same review, laboratory studies were identified in which several older organochlorine chemicals were leached with water in soil columns. The only pesticide found to leach was lindane.

Lindane has also been found in ground water in Israel (*10*). Lindane's estimated retardation factor was the lowest of the older pesticides evaluated. Additionally, the review indicated that the occurrence of ground water contamination resulting from agricultural fertilization was much more prevalent with heavy application of mobile forms of elements such as nitrogen, phosphorus, and chloride.

A 1984 review (*11*) found pesticide leaching from agricultural fields to be more widespread than in 1974. Pesticides currently used are more likely to contaminate ground water than pesticides used in the past. Several general observations can be drawn from the 1984 review:

- Contamination has been limited to shallow, unconfined aquifers consisting of porous sands (DBCP and EDB are exceptions).

- Contamination within the unconfined aquifer has been limited to near field wells (DBCP, EDB, and aldicarb are exceptions).

- Contamination levels found have been in the low parts-per-billion range.

- Contamination has been mainly observed for carbamate, triazine, and low molecular weight halogenated hydrocarbon classes of pesticides (aromatic acids are proposed to be included in this group).

- Contamination is most likely in agricultural areas that have nitrate contamination problems.

- Contamination does not seem to be limited to any one geographical region (pesticide residues have been found in ground water from New York to California).

- Contamination potential is high if the compound does not hydrolyze readily, has a low volatilization potential (Henry's law constant), has a low sorption potential, and possesses a first-order degradation rate in soil of at least 0.02 days^{-1}.

These results for pesticides differ substantially from those found for hazardous wastes. For example, concentrations of hazardous wastes in ground water were found in the hundreds of parts-per-million range, and migration was measured in kilometers rather than meters, as is typical in the near field scenario for pesticides used in agriculture. Several observations can be made concerning these vastly different results for agricultural pesticides:

- The source term (that is, the mass applied) to the land surface is much smaller for pesticides used in agriculture, typically 1.2–2.2 kilograms per hectare as compared to metric tonnes per hectare estimated for hazardous wastes.

- This applied mass is subjected to a variety of soil water processes such as sorption (retardation), transformation or decay, plant removal, and interaction with advection and diffusion–dispersion processes that are conducive to reducing the mass available for infiltration into ground water.

- Analytical capability has greatly increased for pesticides. Nanogram or picogram levels are now commonly identified, whereas a few years ago only milligram levels were obtainable.

Evaluation of Past and Present Pesticides

From our analysis of past and present pesticide use for agriculture, several conclusions can be drawn:

- Pesticide characteristics have changed from the nearly water-insoluble, strongly sorbed, nonmobile compounds of the past to more water-soluble, slightly sorbed, mobile compounds. Part of this change is a result of environmental problems (e.g., bioaccumulation) associated with past compounds; however, pest resistance has also played an important role.

- Ground water contamination by pesticides produced and used through the 1960s probably was not an environmental problem.

- Ground water contamination by pesticides currently produced and used is an environmental problem.

- Low-level residues of pesticides identified in ground water generally may not present acute toxicity problems, but chronic effects will likely be of concern (assuming the toxicity is known).

- Ground water contamination from pesticide use does not necessarily mean that drinking water supplies will be affected unless the ground water is the supply source. Privately owned shallow wells in unconfined aquifers are probably at greater risk than large municipal wells in deeper, confined aquifers.

- Ground water contamination by pesticides involves some risk (pesticides by definition create this risk) and probably will have to be evaluated carefully in view of this risk potential. Pesticides account for 20% of the total agricultural chemical market, and some 3.7×10^5 metric tons of chemicals involving approximately 600 active ingredients are currently applied to agricultural land (*11*).

Carson's prophetic view of the effects of pesticides on the environment contributed greatly to the rapid development of both regulatory and scientific aspects of pesticide evaluation. The regulation of pesticides can be traced back to 1903 when truth-in-labeling laws required products to be efficacious as to label claims. However, the first true pesticide law enacted was the 1947 Federal Insecticide, Fungicide, and Rodenticide Act (FIFRA). This law did not provide for any direct means of evaluating the environmental effects of pesticides; its emphasis was on product efficacy. Shortly after *Silent Spring* was published, scientists were coming to conclusions similar to Carson's, a situation that led to the first request for environmental fate-related data to be submitted for pesticides in 1970. A substantially modified FIFRA was created in 1972 that required indicator test data on efficacy, metabolism, formulation and product chemistry, fish and wildlife effects, toxicology, and environmental fate and transport mechanisms to be provided for each chemical submitted to the U.S. Environmental Protection Agency (EPA) for registration. FIFRA is one of only two congressional laws that require submission of environmentally related data for chemicals manufactured for commercial use.

The occurrence of potentially toxic pesticides in ground water has led to intensive efforts toward environmental risk assessment for chemicals. Ground water contamination from pesticides results from the interaction of the hydrologic cycle and pesticide transport and transformation properties and characteristics. Many different climatic, soil, crop, land management, and chemical variables are inherent in this interaction. *Mathematical models* are currently being used to integrate these variables into a methodology that can provide quantitative indications of the loadings of pesticides to ground water. These models incorporate and simulate the important features associated with pesticide leaching from agricultural fields, that is, quantitative representations of plant–soil–management practices, the hydrologic cycle, and chemical processes and reactions.

The use of models to generate time series data is an accepted technique to derive probability statements about hydrologic events. Similarly, pesticide fate and transport models have been used to estimate probabilities of environmental exposure expressed as frequency distributions of concentrations in surface waters (12, 13). Models that derive probability statements about the mass of pesticides that could contaminate ground water appear to be valuable tools in assigning relative risks to pesticide use.

Future Issues

Significant progress has been achieved in identifying and evaluating the important climatic conditions, soil properties, and chemical properties and characteristics that are conducive to contamination of ground water by pesticides. Agricultural practices are constantly changing, however, to increase the production of food for both American and world needs. Pesticide formulations will evolve to keep abreast of agricultural practices and the changes in pest characteristics.

The number of hectares of fertile productive farmland is decreasing as a result of the sale of farmland for commercial or residential land development and because of the loss of topsoil from erosion. The development and implementation of conservation tillage is being promoted as a means of reducing runoff and erosion losses from farmland. Conservation tillage increases the amount of water that infiltrates into the soil profile. Part of this increased infiltration may be added to the recharge term of the water balance equation. This increase in recharge could increase the contamination of ground water by soluble pesticides.

Conservation tillage has been demonstrated to reduce pesticide loss due to runoff and erosion (14). Unfortunately, there is a lack of monitoring data on pesticide migration to ground water from these practices. A recent model comparison study of conservation tillage (till plant) and conventional tillage (disk harrowing) using three representative pesticides on three agricultural fields indicated conservation tillage increased the potential for pesticide leaching in two of the three fields (15). The effect conservation tillage has on ground water quality clearly warrants further evaluation.

The use of irrigation for agriculture is increasing in some parts of the United States. Chemigation (mixing of pesticides in irrigation water) is being used increasingly to apply pesticides to crops. The use of water-soluble chemicals for chemigation where flood irrigation or chemigation followed by irrigation is used for frost protection may also enhance the migration of soluble pesticides to ground water. The effect of chemigation on ground water quality is also unknown and requires additional study.

Pesticides will continue to be developed in response to new agricultural practices and needs, changes in pest resistance, and technological advances in pesticide chemistry. The development of strategies for the protection of ground water and its beneficial uses must keep pace with this changing technology. Screening methodologies and mathematical models offer one potential cost-effective technique to develop, update, and evaluate such strategies.

Glossary

Advection   The vertical movement of water.

Agronomy   The branch of agriculture dealing with crop production.

Aquifer   A water-bearing bed of permeable rock, sand, or gravel capable of yielding considerable quantities of water to wells or springs. A confined aquifer is one that is confined between two aquitards (confining beds). An unconfined aquifer is one in which the water table forms an upper boundary (not confined).

Hydrologic cycle   The complex sequence of conditions through which water naturally passes from the atmosphere by precipitation on land or water surfaces and ultimately back into the atmosphere by evaporation and animal and plant transpiration.

Nonpolar pesticides   Those that are electrically neutral; that is, they possess no charge.

Octanol–water partition coefficient   A measure of a chemical's ability to dissolve in fat; the higher the coefficient, the higher the ability to dissolve in fat.

Qualitative evaluation   An evaluation that reduces information from a general to a more particular or restricted form.

Tile drain   A hollow unit made of clay buried underneath the soil for constructing a drain.

Time-series data   Output from a mathematical model in terms of mass (grams or milligrams) or concentration per unit time (day, month, or year).

Transformation   A change in composition or structure.

Unsaturated zone   That part of the soil above the water table where soil is not saturated.

Literature Cited

1. Zaki, M. H.; Moran, D.; Harris, D. *Am. J. Public Health* **1982,** 72, 1391–1395.
2. Todd, D. K.; McNulty, D. E. *Polluted Ground Water: A Review of Significant Literature;* U.S. Environmental Protection Agency. U.S. Government Printing Office: Washington, DC, 1974; EPA-600/4-74-001.
3. "Protecting Ground Water: The Hidden Resource" *U.S. Environ. Prot. Agency J.* **1984,** 10, 11–13.
4. Enfield, G. D.; Carsel, R. F.; Cohen, S. Z.; Phan, T.; Walters, D. M. *Ground Water* **1983,** 20, 711–722.
5. Davidson, J. M.; Brusewitz, G. H.; Baker, D. R.; Wood, A. L. *Use of Soil Parameters for Describing Pesticide Movement Through Soils;* U.S. Environmental Protection Agency. U.S. Government Printing Office: Washington, DC, 1975; EPA-600/2-75-009.
6. Stewart, B. A.; Woolhiser, D. A.; Wishmeier, W. H.; Caro, J. H.; Fere, M. H. *Control of Water Pollution from Cropland, Volume II: A Manual for Guideline Development;* U.S. Environmental Protection Agency. U.S. Government Printing Office: Washington, DC, 1976; EPA-600/2-75-026a.
7. Wood, A. L.; Davidson, J. M. *Soil Sci. Am. Proc.* **1975,** 41, 821–825.
8. Selim, H. M.; Davidson, J. M.; Rao, P. S. C. *Soil Sci. Am. Proc.* **1977,** 41, 3–10.
9. Carsel, R. F.; Smith, C. N.; Mulkey, L. A.; Dean, J. D.; Jowise, P. P. *Users Manual for the Pesticide Root Zone Model: Release I;* U.S. Environmental Protection Agency. U.S. Government Printing Office: Washington, DC, 1984; EPA-600/3-84-109.
10. Lahav, N.; Kahanovitch, Y. *Water, Air, Soil Pollut.* **1974,** 3, 253–259.
11. Cohen, S. Z.; Creeger, S. M.; Carsel, R. F.; Enfield, C. G. In *Potential Pesticide Contamination of Ground Water from Agricultural Uses;* Krueger, R. F.; Seiber, J. N., Eds.; ACS Symposium Series No. 225; American Chemical Society: Washington, DC, 1984; pp 298–325.
12. Mulkey, L. A.; Falco, J. W. In *Methodology for Predicting Exposure and Fate of Pesticides in Aquatic Environment;* Scallen, F. W.; Bailey, G. W., Eds.; Agricultural Management and Water Quality, Iowa State University Press: Ames, Iowa, 1983; pp 250–266.
13. Onishi, Y.; Brown, S. M.; Parkhurst, M. A.; Wise, S. E.; Walters, W. H. *Methodology for Overland Flow and Instream Migration and Risk Assessment of Pesticides;* U.S. Environmental Protection Agency. U.S. Government Printing Office: Washington, DC, 1982; EPA-600/3-82-84.
14. Baker, J. L.; Laflen, J. M.; Johnson, H. P. *Trans. ASAE* **1978,** 5, 886–892.
15. Carsel, R. F.; Smith, C. N.; Parrish, R. S. *Modeling Differences Between Conservation and Conventional Tillage on Pesticide Leaching;* U.S. Environmental Protection Agency: Athens, GA; in review.

6 ᴄᴏ Impact of Pesticides on Bird Populations

Russell J. Hall

Rachel Carson and *Silent Spring* are intimately connected with the studies on the impact of pesticides on bird populations conducted by the U.S. Fish and Wildlife Service. When I began investigating that connection, I was certain that it indeed existed and that knowledge of it would provide much insight on the impact of pesticides in the years before and after *Silent Spring*. What I learned was surprising, but not disappointing.

Patuxent Wildlife Research Center in Laurel, Maryland, is the oldest and largest research facility of the U.S. Fish and Wildlife Service. Since 1945 it has had national responsibility for investigating and predicting the effects of environmental chemicals on wildlife populations. Although figures have not been compiled, most likely a significant portion of research on the chemical effects on wildlife published in the United States has come from Patuxent or was done under its auspices. Approximately 100 full-time employees currently work in Patuxent's contaminant research program, and they work

0980–4/87/0085$07.75/0 © 1987 American Chemical Society

closely with biologists and other employees assigned to other parts of the agency. The examples I will be using are almost exclusively based on work done at Patuxent and at its field stations, although a huge volume of important work has been done elsewhere.

This chapter examines where *Silent Spring* fits into the 40 years of work on the effects of pesticides on wildlife conducted at Patuxent, and it evaluates how this work, past and present, relates to the major assertions of *Silent Spring*.

Early Studies

Research on pesticides began at Patuxent soon after the beginning of the modern pesticide era and the introduction of DDT in 1943. Our first results of experiments with DDT bear publication dates of 1946. Five papers were published that year, one relating the results of some acute toxicity tests (1) and four reporting the effects of two different experimental spray programs on wildlife populations (2-5). The toxicity of DDT to wildlife was demonstrated as were some deaths of free-living wildlife from sprays at high rates of application (5 pounds per acre), but attempts to show significant population declines generally failed. The fifth 1946 publication summarized (6) the published studies and some other studies that began as early as 1944 and, in a notable set of recommendations, cautioned users of the potential dangers of DDT to wildlife. Briefly summarized, the recommendations are as follows:

- Use DDT for the control of an insect pest only after weighing the value of such control against the harm that will be done to beneficial forms of life.

- Use...less than 2 pounds per acre to avoid damage to birds, amphibians, and mammals in forest areas.

- Use DDT only where it is needed.

- In forest-pest control, wherever feasible, leave strips untreated at the first application to serve as undisturbed sanctuaries for wildlife....

- In the control of early appearing insect pests, apply DDT, if possible, just before the emergence of leaves and the main spring migration of birds.... Adjust crop applications and mosquito applications so far as is possible to avoid the nesting period.

- ...Avoid as far as possible direct applications to streams, lakes, and coastal bays.

- Whenever DDT is used, make careful before and after observations of mammals, birds, fishes, and other wildlife.

As early as 3 years after the introduction of DDT, at a time when it was understandably hailed as a great benefit to mankind, a consciousness of the potentially destructive effects of the chemical and a mechanism to measure and predict such effects through research developed within the U.S. Fish and Wildlife Service.

The evidence of this earlier work disputes the popular impression that Carson awakened a sleeping bureaucracy to investigate heretofore unrecognized threats. To the contrary, *Silent Spring* was an outgrowth of a consciousness that developed simultaneously within the scientific community and the U.S. Fish and Wildlife Service, particularly at Patuxent, during 16 years of research on pesticides and wildlife.

Research in the 1950s

Research conducted by Patuxent scientists during the next few years focused on censuses and other observations made on experimental or large-scale operational DDT spray programs. Studies were done on the Patuxent Refuge (7, 8); at the Beltsville Agricultural Research Center (9); in forest spray programs in Idaho and Wyoming; at orchard sites in Maryland and Georgia; in an area of Texas where DDT was used for tick control (10); in Princeton, New Jersey, where DDT was used to control Dutch elm disease (11); in a marsh area of New Jersey subjected to mosquito control (12); and elsewhere as opportunities arose. Scientists realized early that laboratory studies could do little more than establish that insecticides are toxic to various wildlife species; the studies' applicability to the real world was always subject to question. For this reason, a conscious effort was made to concentrate research in the field, where DDT was actually in use.

By the early 1950s, it was well established that dead birds were commonplace in fields sprayed with more than 5 pounds per acre of DDT. Further, observations during periods as long as 5 years demonstrated that as little as 2 pounds per acre applied annually could reduce the density and reproduction of forest birds (7). Indications of chronic toxicity from field studies led to laboratory tests of pesticides other than DDT to provide a predictive basis for evaluating their impacts on wildlife.

In the mid-1950s, laboratory research conducted at Patuxent became much more toxicologically sophisticated. Until then, dead birds found after sprayings could not be assumed to have died as a direct result of the chemicals. Once the ability to detect and quantitate DDT in tissues was developed, the problem became more complex. In a 1953 study (13), for example, the DDT content of birds found dead in a treated area ranged from a trace to 77 parts per million in breast muscle. Experimental dosing studies conducted later confirmed that the amounts of DDT stored in tissues were not proportional to the amount fed, to the duration of exposure, or to the survival time. Despite much effort, the exact relationship between the amounts of pesticide found in tissues and the toxic effect was not elucidated for some time. Advances in analytical chemistry that permitted the more precise measurement of pesticides in animal tissues were made during this period and were invaluable in the eventual effort to establish the necessary cause–effect relationships.

Another advance of the mid-1950s was the discovery (14) that the pesticides DDT, DDD, endrin, aldrin, and dieldrin, incorporated into the diets of pheasants and quail, lowered reproductive success without other apparent effects. Although relatively primitive in comparison to later studies, these tests provided results that influenced future thinking on the evaluation of the impacts of pesticides on wildlife.

The proposal to eradicate the imported fire ant by using heptachlor and dieldrin in 1957–1958 was important for the future of pesticide–wildlife relationships. The massive mortality of wildlife resulting from this program and the efforts mounted by the U.S. Fish and Wildlife Service and others to document these effects increased concerns about the safety of pesticide use. The signing on August 1, 1958 of the Magnuson–Metcalf Bill, which became Public Law 85-582, was another signal event. Most previous pesticide studies were conducted cooperatively with the U.S. Department of Agriculture, but the U.S. Fish and Wildlife Service now had a specific mandate and enhanced funding to "undertake comprehensive continuing studies on the effects of insecticides, herbicides, fungicides, and pesticides upon the fish and wildlife resources of the United States...." Studies began on the acute and chronic toxicity of 200 basic pesticidal chemicals, chemical analyses were conducted to determine suspected cases of pesticide poisoning, field appraisals of large-scale pest control operations were carried out, and compilation and dissemination of findings began so hazards to "desirable forms of life" might be minimized.

Although most benefits of Public Law 85-582 would not be seen for a few years, its requirement that past findings be disseminated may have encouraged the 1960 publication of a summary of results obtained through 1959 known as *Fish and Wildlife Service Circular 84 (15)*. Preliminary findings from most recent efforts, including the imported fire ant control studies, new studies of DDT for Dutch elm disease control, and reports of chemical analyses of wildlife collected in spray areas, were presented briefly, and the results of studies conducted elsewhere were summarized. Recommendations for safeguarding wildlife were included at the end of the report. These recommendations were surprisingly similar to those made in a 1946 report, but included the following extensions:

- More attention should be given to developing chemicals specifically toxic to only one particular group of animals.

- Biological methods of control also should have more study.

- Other promising control methods include planting and harvesting at particular times, proper fertilization and rotation of crops, destruction of insect wintering quarters, and manipulations of water levels.

• ...The development of varieties of plants and animals that are resistant to troublesome insects and disease (may hold) the greatest promise of all from the long-term standpoint.

Circular 84 was the last really significant U.S. Fish and Wildlife Service report before the publication of *Silent Spring*. Circular 84 reads as if it might have been an intellectual blueprint for certain parts of *Silent Spring*. The report summarized, albeit briefly, the major scientific data on wildlife effects that would later support the wildlife portions of the book. Also, Circular 84 saw pesticides as more than the unmitigated boon scientists and laymen alike believed them to be; hence, the report recommended cautious use and further study. Carson might have influenced the scientific community and the U.S. Fish and Wildlife Service in later years, but undoubtedly the visionary scientists who cared about pesticides' impacts on wildlife were vital to the creation of *Silent Spring*. At the time of *Silent Spring*'s publication, Patuxent biologists and others in the field understandably felt that Carson was carrying their banner.

Research Contemporary with *Silent Spring*

In 1959, the ceiling on funds in support of Public Law 85–582 was raised from $280,000 to $2,565,000. The actual amount appropriated for fish and wildlife research (including research on commercial fisheries) was $852,000 in fiscal year 1963. New legislation was proposed so that pesticide regulations would recognize the value of fish and wildlife. At the dedication of a new chemistry–physiology building at Patuxent in 1963, Secretary of the Interior Stewart Udall acknowledged that "We owe much to Rachel Carson." Wildlife research programs grew not only at Patuxent, but also at the Denver Wildlife Research Center and at several of the Cooperative Wildlife Research Units during this time.

Despite numerous changes in organization and minor changes in scope, U.S. Fish and Wildlife Service research on the effects of pesticides on wildlife has remained relatively active and well supported up to the present. Most of the work done during 1959–1964 was published in the "gray literature", a series of reports summarizing findings on an annual basis (*16–19*).

Silent Spring appears to have increased acceptance of papers on pesticides and wildlife in national scientific journals. From 1964 to 1984, 429 published papers on pesticides and wildlife came from Patuxent alone.

One of several experimental pen complexes at Patuxent Wildlife Research Center. Large numbers of spacious pens are required to determine the effects of low levels of environmental contaminants on the reproduction of captive wildlife.

Major Research Findings After 1962

The Meaning of Pesticide Residues

To diagnose the death of birds from pesticides, it is necessary to have in hand not only a dead bird and the ability to detect residues of the pesticide in it but also interpretive data to relate the measured residues to death. Initial attempts to correlate death and residues were not satisfactory. Residue levels found in dead birds varied greatly, even when the birds were killed by the same chemical under similar circumstances. A long series of experiments on the kinetics of pesticide residues in birds indicated by 1964 that residues of DDT in the brain were reliably related to lethality (20). Additional work related lethal brain levels of DDT to fat cycling or depletion in the body, extended these findings to other species, and, ultimately, reported diagnostic lethality levels for a number of other organochlorines. Lethal residue concentrations were determined for DDE, dieldrin, endrin, polychlorinated biphenyl mixtures, heptachlor epoxide, components and metabolites of chlordane, and mirex. Stickel (21) summarized most of this early work.

These developments permitted not only reliable diagnoses of death from organochlorines on animals in the field, but also helped us evaluate the significance of residues found in apparently healthy animals. These discoveries confirmed and extended the results of Silent Spring (22) by revealing mortality from stored pesticide residues.

Synergism, Additivity, or Independence. Silent Spring suggested that sublethal or seemingly harmless residues of many chemicals could interact in the body to produce unexpected or enhanced effects. Although studies done at Patuxent have not answered all the important questions in this area, preliminary findings are available. Pheasants and quail were fed 13 pairs of chemicals, some chosen because of their synergism in other animal groups. Two of the pairs showed moderate synergism, but the remaining pairs appeared to be additive in effect (23). The quail were fed a formulation of methylmercury and were later dosed with parathion. Synergism was shown, but both chemicals are known inhibitors of acetylcholinesterase (24), so synergism could be expected. Dieldrin residues increased when DDT was added to the quails' diet (25), and chlordane and endrin, which are structurally similar to dieldrin, were found to be additive in their toxicity (26). Chlordane pretreatment seemed to reduce sensitivity to parathion, whereas DDE pretreatment enhanced parathion toxicity (27). Additivity or synergism was suggested in studies of chlordane components and metabolites (28), but another study (29) indicated that dieldrin and endrin residues in the brain were not additive in producing mortality, even though dieldrin and endrin are stereoisomers and would be expected to have similar modes of action.

These results tend to allay fears expressed in Silent Spring (30) that apparently harmless quantities of a variety of chemicals in the body may combine to produce adverse effects. Although the full effects of chemicals in combination have not yet been revealed, synergism is relatively rare and probably limited to chemicals with the same target systems. Additivity may be more common, but it cannot be assumed in all cases. Investigators can probably diagnose death from pesticides in 90% or more of all cases by assuming that pesticides in the body act independently of one another.

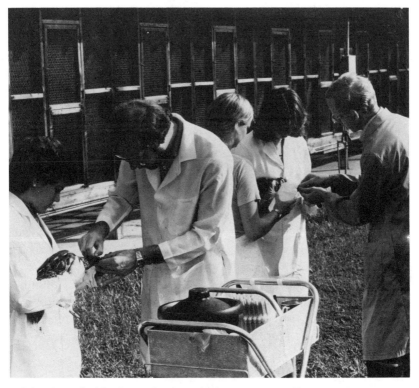

Scientists take blood samples from ducks experimentally exposed to a pesticide. Measurement of certain blood chemicals can serve as an indication of subtle effects of pesticide exposure.

Global Distribution of Residues of Persistent Pesticides. Investigative monitoring of pesticide residues in wild animals began as soon as chemical methodology made such large-scale efforts possible. Surveys of woodcock wings and black duck eggs in the early 1960s and continuing analysis of the carcasses of bald eagles showed an almost universal distribution of DDT residues. Animals in Antarctica and from areas never sprayed also contained traces of DDT or its metabolites (*31, 32*).

A formal monitoring program was established as a result of these revelations. DDT residues were found in all 66 eagles collected in 1964-1965 (*32*). In 1969 polychlorinated biphenyls (PCBs) were detected in bald eagles by Patuxent chemists (*33*); PCBs were detected earlier in tissues of falcons at another laboratory (*34*). All woodcock wings sampled in 11 states in 1970-1971 had residues of DDT, PCBs, and mirex (*35*). Marine birds (*36*), marine mammals (*37*), sea turtles (*38*), and crocodiles (*39*), among many other animals, were contaminated.

Monitoring of pesticide levels in duck wings (wings of game ducks contributed by hunters) began in 1965. In 1967, Patuxent began monitoring pesticide levels in the carcasses of starlings. Most pesticide residues have continually decreased since the early 1970s; DDE in starlings decreased from a nationwide average of 1.6 parts per million in 1967-1968 to 0.17 parts per million in 1979. DDE in duck wings from the Pacific Flyway decreased from nearly 1 part per million in 1969 to 0.35 parts per million in 1979. Other pesticide residues also declined (*40-42*). Declines in residues in these indicator species largely paralleled those seen in investigative monitoring of a variety of species, including bald eagles.

The confirmation of global distribution of pesticide residues supports preliminary information reported in *Silent Spring* (*43*). The declines are, of course, a favorable result of the restriction or banning of persistent pesticides.

Results of Toxicity Screening

Screening of pesticides for indications of possible adverse effects on wildlife began relatively early and developed along three principal lines. The Denver Wildlife Research Center relied primarily on acute toxicity tests in which representative wildlife species were given pesticides in single oral doses. At Patuxent, subacute testing predominated; quails, pheasants, or mallard ducks received pesticides mixed in their diets for 5-day test periods. Later at Patuxent, quail or mallard eggs were used to detect toxic or teratogenic effects on embryos. More than 350 pesticides, components of pesticides, metabolites, or formulations have been tested, most on multiple species and with more than one of the procedures. The results have been summarized in a routinely updated series of publications (44–48).

Pesticide screening is used to detect compounds that might produce severe effects in the field and to flag them for further testing. In some cases, screening has accurately predicted hazardous pesticides. However, only about 25 of the 350 compounds tested have been implicated in wildlife problems in actual use.

Based on their low toxicity, their use patterns, or their records of safe use, most chemicals tested at Patuxent have appeared to be relatively safe for wildlife; consequently they have not been tested further. Those selected for further testing are suspected of having adverse effects on wildlife. Wildlife does not seem to be seriously threatened by most chemicals tested. Extensive testing of methoxychlor, lindane, malathion, and carbaryl has failed to indicate any direct adverse effects on wildlife, although each chemical could have indirect effects by reduction of prey organisms. Preliminary work on synthetic pyrethroids and fragmentary work on a great number of cholinesterase inhibitors has given us little reason to suspect that they cause direct wildlife hazards, although some long-term or secondary effects have been documented by more thorough investigations. We could find few direct hazards to birds from mirex or toxaphene, although registration of each was later canceled because of human health considerations or toxicity to aquatic life.

The results of this large-scale screening restrict some contentions of *Silent Spring* (49) that may have left readers believing that nearly all synthetic pesticides might be implicated in wildlife problems.

Shell Thinning

Patuxent scientists were aware of the reproductive problems faced by large predatory birds and were reasonably certain that DDT was involved. Ospreys (19) and bald eagles (50) were severely contaminated by pesticides and had massive reproduction problems. Experiments were begun with the new breeding colony of kestrels (small falcons) in an attempt to elucidate the problem. Before the results of these experiments were available, a British scientist reported (51) that eggshell weight in European birds of prey had decreased since the introduction of modern pesticides. The work was soon repeated in the United States by a group at the University of Wisconsin under contract to Patuxent (52). Thus, shell thinning was shown by correlative data to be a major mechanism of pesticide impact on wildlife populations.

Shell thinning was soon demonstrated in captive kestrels. The birds, fed a combination of dieldrin and DDT, had reduced shell thickness and the attendant breakage and disappearance of eggs seen earlier in free-living raptors (birds of prey) (53). Experimental proof was thus obtained that the effects observed in nature were indeed produced by organochlorine pesticides. Impaired reproduction and shell thinning were subsequently demonstrated with DDE in the diets of mallard ducks (54) and black ducks (55). Additional studies (56) showed that DDE alone could thin eggshells in kestrels as well.

A captive American kestrel. A large breeding colony of these birds at Patuxent permits experiments that represent the response of breeding birds of prey to chemical contamination.

Defining the Limits of the Problem. At first, it was feared that all birds were susceptible to shell thinning and that all persistent pesticides might cause it. Early work on this subject at Patuxent sought to define the limits of the problem. Sensitivity to shell thinning varied greatly among different groups of birds observed; sensitivity levels were based on both ecological and physiological factors. In the many species susceptibile to shell thinning, effects on reproductive success seemed to be restricted primarily to some raptors and fish-eating birds (21). Field data (57) and numerous laboratory studies indicated that DDE is responsible for the overwhelming majority of shell thinning observed.

The discovery of shell thinning revealed a mechanism of pesticide impacts on wildlife that *Silent Spring* did not anticipate (58); the book focused on the apparent effects of DDT on gamete production. The finding that shell thinning seriously involved only one pesticide and relatively few species restricted fears that were expressed soon after the discovery of the problem.

South Carolina Brown Pelicans. Once shell thinning was discovered, efforts were made to demonstrate the relationship between this phenomenon and the decline of bird populations. In 1969, Patuxent began studies on South Carolina brown pelicans. These pelicans had eggshells significantly thinner than those in Florida; the thinning was close to that seen in declining populations of raptors. The number of breeding pairs in the colonies studied declined from more than 5000 in the early 1960s to 1250 in 1969. In that year, residues of DDE in eggs averaged 4.65 parts per million and, from 1969 through 1972, the number of young fledged per nest averaged approximately 0.8, far below the 1.2–2.5 required to maintain the population.

However, as residues in the environment began to decline, there was a turnabout in fledgings. By 1975, DDE in eggs declined to close to 1 part per million, and shell thickness increased from a low of 0.46 millimeters in 1970 to 0.52 millimeters in 1977. The number of nests increased to 3300 by 1977 and the number of young fledged per nest was 1.4 (59). The decline of this population and its recovery in response to the phaseout of DDT provided documentation both of the

An adult and nestling brown pelicans on the breeding grounds, Cape Romain, South Carolina.

deleterious effects of the pesticide and the benefits from eliminating the principal variable—environmental DDT. Ospreys, bald eagles, and peregrine falcons seem to be experiencing similar recoveries but at a much slower rate that is more difficult to observe.

Die-offs of Free-Living Birds

Wildlife managers, law enforcement personnel, and others commonly ask Patuxent to investigate die-offs of wildlife in the field when pesticides are suspected. Our ability to diagnose the cause of death in such cases was enhanced by the routine diagnostic use of brain cholinesterase assays as indicators of exposure or death from cholinesterase-inhibiting chemicals (*60, 61*). In most cases, we chemically analyze stomach contents to

determine the specific chemicals involved after a lethal exposure to a cholinesterase inhibitor. Thirteen organophosphates
and one carbamate pesticide were implicated in poisoning
incidents by our studies; those accounting for more than one
episode each are parathion, Diazinon, monocrotophos, phorate,
dicrotophos, and carbofuran. Additional die-offs were summarized recently (62).

The number of die-offs referred to Patuxent has remained
fairly constant in recent years; 18 die-offs were investigated in
the first 10 months of fiscal year 1984. Die-offs that come to our
attention are undoubtedly a small fraction of the total number
that occur. Cases that attract attention and are referred to our
laboratory are generally those involving large, conspicuous
birds or large numbers of birds found dead in one place or both;
less obvious cases almost always go unnoticed. Our continued
involvement with die-offs indicates that birds continue to be
killed by pesticides despite an elaborate system of safeguards.

These studies confirm and extend *Silent Spring*'s reports of
die-offs. Fortunately, however, relatively few chemicals are
involved.

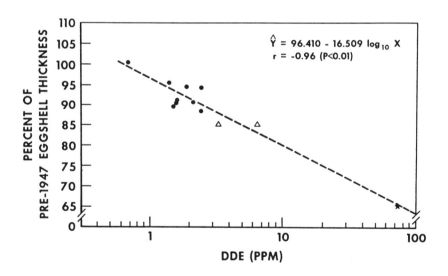

Graph depicting the relationship of DDE (a major breakdown product of
DDT) and eggshell thickness in brown pelicans, based on field data from
Florida (●), South Carolina (△), and California (★).

Crushed and normal eggs in a Caspian tern nest. DDT residues and consequent shell thinning have been related to such breakage.

Long-term, Delayed, or Secondary Effects of Organophosphates and Carbamates

Since 1979 Patuxent scientists have conducted approximately 25 major studies investigating possible indirect effects of cholinesterase-inhibiting pesticides on wildlife. Some of the secondary poisoning studies have suggested (63) or confirmed (64) a hazard in the field. Exposures to organophosphates or carbamates have resulted in changes in behavior (65), in hormones and in cold tolerance (66), in salt gland function (67), in growth (68), and in embryonic development (69). No study has indicated the types or magnitude of problems seen with the organochlorines widely used in the past, although some indications of possible threat require further follow-up.

The results of these studies have vindicated many compounds that were suspected of causing significant wildlife problems. *Silent Spring's* references to potential problems resulting from so-called safe pesticides (70) led many to expect that research on them would eventually reveal wildlife problems in the field.

Appearance of Organochlorine Hot Spots

Although residues of most persistent pesticides have declined in most places, residues have remained constant or even increased in certain areas (71). Heptachlor residues in Oregon and endrin residues in the upper Midwest and in orchards in Washington state remain a problem. The greatest hot-spot problem involves DDT, the oldest of the modern pesticides.

DDT metabolites are present at dangerous levels in large areas of the West and Southwest. It has been suggested that DDT residues are picked up by migratory species that winter in parts of Latin America where, it is presumed, pesticide use is unrestricted. Another hypothesis is that DDT compounds are released as an impurity by the pesticide dicofol or are produced in the environment by the metabolic conversion of dicofol. A third hypothesis is that much residue of DDT remains in the environment from heavy historical use, and a fourth is that illegally imported DDT is currently in widespread use. Investigations so far point toward the last two explanations as the most likely origins of the observed DDT compounds. More than 12 years after its removal from the market, DDT continues to have major impacts on wildlife. The continued problems with organochlorines confirm the observations of *Silent Spring* regarding the long-term, residual effects of the "old" pesticides.

Wildlife Threats from Chemicals Other Than Pesticides

Beginning in the 1960s, Patuxent's efforts were partly diverted from pesticides to toxicological studies of lead shot. Later, other metals and industrial pollutants were studied. The effects of petroleum hydrocarbons in the environment were assessed in the mid-1970s. Today, half or more of the environmental contaminant research at Patuxent is devoted to contaminants other than pesticides. The most ominous contaminant that has come into recent focus is a byproduct of an agricultural practice in California. Selenium, released through subsurface drainage of irrigation water, accumulates in marshes and bays where it impairs the reproduction of water birds and kills chicks and adults. Selenium released in this manner may be "the DDT of the future" in terms of its effect on wildlife populations.

In Summary

Despite various allegations by her critics, Carson was right; pesticides were massively and needlessly killing our wildlife, and even bringing some species close to extinction. In 40 years of work at Patuxent, we have tried to separate fact from fiction and real threats from unproven ones. To the extent that we and other investigators have succeeded, we have helped protect our wildlife without eliminating chemical control of pests.

Work Left Undone

Despite assurance that the major threats to wildlife resources from pesticides have been identified and contained, several assertions of *Silent Spring* and questions arising out of work at Patuxent and elsewhere have not been adequately addressed. In some instances Patuxent scientists tried to find answers but failed, in other instances they have not yet found the time to ask the questions in a suitable way, and in still other cases, they may have lacked the necessary insight to see the true nature of the problem. Some of the failings and unfinished agendas follow.

Significance of Die-offs Unknown

The frequency or magnitude of pesticide-caused die-offs cannot yet be projected on a national, regional, or even local basis. As mentioned, the samples Patuxent diagnoses are highly biased and are likely to represent but a small fraction of the total mortality. Studies have not been performed that would permit an estimate of the real losses of wildlife in the field.

A recent photograph taken at the Wheeler National Wildlife Refuge, which is contaminated by the nearby outfall of an old manufacturing plant. It illustrates that some problems with the "old" pesticides have persisted into the 1980s.

Interaction of Contaminants with Environmental Stresses

To ensure repeatability and comparability, experiments are usually performed under optimal conditions for the test animals. Yet maximum susceptibility to pesticides may occur at times of unusual natural stress. Disease, food deprivation, predation, and competition are all factors that cause mortality in the field. Studies are needed to indicate what role, if any, the effects of pesticide exposure have in increasing vulnerability to these other factors.

Population Effects of Pesticide Sprays

Although early work with DDT assessments had limited success, scientists still have little idea of the effects of large-scale pesticide use on bird populations other than colonial species like pelicans or birds with sparse populations like eagles and ospreys. Normal short-term surveys have proven unreliable, and the establishment of long-term population trends over wide areas is costly and difficult. Modern models for population estimation need to be applied to the problem; if effective, this approach will help in the design of controlled experimental studies that have the statistical power to detect biologically significant differences among affected groups.

Effects in Other-Than-Prime Wildlife Habitats

Most of Patuxent's fieldwork has been concentrated in prime wildlife habitats such as coastal islands, forests, rangelands, and National Wildlife Refuges. Relatively few studies have been done on the agricultural and urban areas that receive the greatest pesticide use. Because the most severe effects are localized, observers may be missing most of the effects. For example, heavy wildlife mortality from endrin was documented only after we began looking in orchards where it was used for rodent control (72). Also, potential wildlife reductions in areas of heavy pesticide use such as the Central Valley of California need to be further investigated.

Interactions of Chemicals

More basic work needs to be done on chemical interactions. Although harmful interactions seem to be rare, much more must be known because one potent interaction could have far-reaching effects.

Environmental Effects

Effects of pesticide use on ecosystems have scarcely been investigated. Reduction of the food supply, disruption of delicate predator–prey relationships, alteration of the habitat through herbicide use, and a variety of similar changes could all affect the distribution and abundance of wildlife. Studies of these effects would be difficult and could be inconclusive unless basic methodologies for population estimation are developed and appropriately applied in the field.

The Organochlorine Mentality

A history of successful research on organochlorines may have made scientists less able to assess and predict the effects of other classes of pesticides such as the so-called biorational kinds. Past research has emphasized using species of predatory birds for tests, observing lethality or reproductive effects as end points, using oral routes of exposure, and relying on chemical analysis to ascertain exposure. The biases resulting from these approaches may lead researchers to overlook the real hazards of the next significant threat as surely as a fixation on direct toxicity may have caused researchers in the past to at first overlook the problem of thin eggshells.

Glossary

Acute toxicity   A measure of the effect in an organism of a poison when it is administered directly, usually as a relatively large oral dose expected to act quickly.

Additive   The tendency of chemicals acting in combination to produce effects equal to the sum of the effects of each acting alone.

Biorational pesticides   Pesticides that are not poisons in the traditional sense, but tend to control pests by interfering in some specific part of their life cycles (e.g., pheromones or chemical signalling compounds that may distract male and female insects from locating each other and successfully mating).

Brain cholinesterase assay   A test that measures the activity of the enzyme cholinesterase in the brain. It has proven to be diagnostic of poisoning by certain pesticides.

Carbamates   Synthetic organic pesticides that are similar in action to organophosphates.

Cholinesterase-inhibiting compounds   Pesticides that achieve their toxic effect by interfering with the enzyme cholinesterase, an essential component in the process of transmitting impulses between nerve endings.

Chronic toxicity   A measure of the effects of a poison on an organism exposed to small amounts of it for an extended period of time.

Die-off   An episode in which wildlife are found dead in the field, from chemical contamination or other causes.

Direct toxicity   Tending to affect organisms by killing them outright, rather than indirectly, by affecting reproductive or other life processes of longer-term significance.

End points   In a test or experiment, the criteria one uses for determining whether or not one has a result. In a test of the toxicity of a chemical, one might conclude that environmental levels of the chemical were safe because they didn't kill any animals; the same levels might be determined to be harmful because production of offspring was reduced. The test was the same, assuming the experiment would not have stopped, but the conclusions differed because different end points were used.

Gamete   Reproductive product of sexually reproducing organisms; sperms and eggs in animals.

Indicator species   The idea that certain species that are easily tested or sampled (starlings, for example) can provide an estimate of the vulnerability or contamination of similar animals in the same environments.

Mechanism   The specific way in which a pesticide achieves its desired effect; how it works in killing pest animals.

Organochlorines   A class of chemical compounds produced by the addition of chlorine atoms to hydrocarbons. Many of them (e.g., DDT, dieldrin, and endrin) had insecticidal properties and became the most successful of the early synthetic insecticides.

Organophosphates   Synthetic organic pesticides that contain phosphorus and tend to attack the enzyme cholinesterase at the nerve endings.

Petroleum hydrocarbons   Any of the great number of chemicals based on carbon, hydrogen, and oxygen that are components of crude oil and its refined products.

Pyrethroids   Synthetic organic insecticides that mimic the structure and activity of pyrethrum, a natural insecticide produced by plants.

Secondary effects   Effects not normally expected from a knowledge of the primary action of a pesticide. Eggshell thinning, for example, would not be predicted from the knowledge that DDT is a poison that is believed to act on nerves.

Stereoisomers   Chemical compounds of identical structure and composition except for the alignment of chemical bond angles.

Subacute testing   Toxicity tests in which an environmental contaminant is administered to a subject in its feed, so as to mimic environmental exposure. In *acute testing*, the substance alone is administered to subjects.

Synergism   The tendency of chemicals acting in combination to produce effects greater than the sum of the effects of the individual chemicals.

Target systems   Physiological processes and components (e.g., biochemical transmission of impulses between nerves) that are inhibited or damaged by particular poisons.

Teratogenic   The tendency to cause the production of developmental abnormalities in fetuses, commonly called birth defects in mammals.

Acknowledgments

I thank Nancy C. Coon for several discussions that helped me to prepare this paper, Nancy Bushby and Linda Garrett for obtaining printed materials, and Marilyn Whitehead, Diana Mullis, and Sharon Fox for helping to prepare the drafts. Helpful comments were provided by Donald R. Clark, Jr., and Charles J. Henny of Patuxent and Shirley A. Briggs of the Rachel Carson Council, Inc.

Literature Cited

1. Coburn, D. R.; Treichler, R. *J. Wildl. Manage.* **1946,** *10,* 208–216.
2. Hotchkiss, N.; Pough, R. H. *J. Wildl. Manage.* **1946,** *10,* 202–207.
3. Mitchell, R. T. *J. Wildl. Manage.* **1946,** *10,* 192–194.
4. Stewart, R. E.; Cope, J. B.; Robbins, G. S.; Brainerd, J. W. *J. Wildl. Manage.* **1946,** *10,* 195–201.
5. Stickel, L. F. *J. Wildl. Manage.* **1946,** *10,* 216–218.
6. Cottam, C.; Higgins, E. *U.S. Fish and Wildlife Service Circular No. 11;* U.S. Government Printing Office: Washington, DC, 1946.
7. Robbins, C. S.; Springer, P. F.; Webster, C. G.; *J. Wildl. Manage.* **1951,** *15,* 213–216.
8. Stickel, L. F. *J. Wildl. Manage.* **1951,** *15,* 161–164.
9. Robbins, C. S.; Stewart, R. E. *J. Wildl. Manage.* **1949,** *13,* 11–16.
10. George, J. L.; Stickel, W. H. *Am. Midl. Nat.* **1949,** *42,* 228–237.
11. Benton, A. H. *J. Wildl. Manage.* **1951,** *15,* 20–27.
12. Springer, P. F.; Webster, J. R. *U.S. Fish and Wildlife Service Special Scientific Report: Wildlife No. 10;* U.S. Government Printing Office: Washington, DC, 1951.
13. DeWitt, J. B.; Derby, J. V., Jr.; Mangan, G. F., Jr. *J. Am. Pharm. Assoc.* **1955,** *44,* 22–24.
14. DeWitt, J. B. *Agric. Food Chem.* **1955,** *3,* 672–676.

15. DeWitt, J. B.; George, J. L. *U.S. Fish and Wildlife Service Circular No. 84;* U.S. Government Printing Office: Washington, DC, 1960.
16. DeWitt, J. B.; Crabtree, D. G.; Finley, R. B.; George, J. L. *U.S. Fish and Wildlife Service Circular No. 143;* U.S. Government Printing Office: Washington, DC, 1962; pp 4–15.
17. DeWitt, J. B.; Stickel, W. H.; Springer, P. F. *U.S. Fish and Wildlife Service Circular No. 167;* U.S. Government Printing Office: Washington, DC, 1963; pp 74–96.
18. Stickel, L. *U.S. Fish and Wildlife Service Circular No. 199;* U.S. Government Printing Office: Washington, DC, 1964; pp 77–115.
19. Stickel, L. F.; Heath, R. G. *U.S. Fish and Wildlife Service Circular No. 226;* U.S. Government Printing Office: Washington, DC, 1965; pp 3–30.
20. Bernard, R. F. *Publ. Mus. Mich. State Univ.* **1963,** *2,* 155–192.
21. Stickel, L. F. In *Environmental Pollution by Pesticides;* Edwards, C. A., Ed.; Plenum Press: London and New York, 1973; pp 254–312.
22. Carson, R. *Silent Spring;* Houghton Mifflin: Boston, 1962; pp 188–191.
23. Kreitzer, J. F.; Spann, J. W. *Bull. Environ. Contam. Toxicol.* **1973,** *9,* 250–256.
24. Dieter, M. P.; Ludke, J. L. *Bull. Environ. Contam. Toxicol.* **1975,** *13,* 257–262.
25. Ludke, J. L. *Bull. Environ. Contam. Toxicol.* **1974,** *11,* 407–414.
26. Ludke, J. L. *Bull. Environ. Contam. Toxicol.* **1976,** *16,* 253–260.
27. Ludke, J. L. *Pestic. Biochem. Physiol.* **1976,** *7,* 28–33.
28. Stickel, L. F.; Stickel, W. H.; McArthur, R. D.; Hughes, D. L. In *Toxicology and Occupational Medicine. Proceedings of the Tenth Inter-American Conference on Toxicology and Occupational Medicine, Key Biscayne (Miami) Florida;* Deichmann, W. B., Organizer; Elsevier/North Holland: New York, 1979; pp 387–396.
29. *Fisheries and Wildlife Research 1977;* Scott, T. G.; Schultz, H. C.; Eschmeyer, P. H., Eds.; U.S. Government Printing Office: Washington, DC, 1978; p 29.
30. Carson, R. *Silent Spring;* Houghton Mifflin: Boston, 1962; pp 31–32.
31. Dustman, E. H. In *Scientific Aspects of Pest Control,* Publ. No. 1402; National Academy of Sciences-National Research Council: Washington, DC, 1966; pp 343–351.
32. Reichel, W. L.; Cromartie, E.; Lamont, T. G.; Mulhern, B. M. *Pestic. Monit. J.* **1969,** *3,* 142–144.
33. Bagley G. E.; Reichel, W. L.; Cromartie, E. *J. Assoc. Off. Anal. Chem.* **1970,** *53,* 251–261.
34. Risebrough, R. W.; Reiche, P.; Herman, S. G.; Peakall, D. B.; Kirven, M. N. *Nature* **1968,** *220,* 1098–1102.
35. McLane, M. A. R.; Stickel, L. F.; Clark, E. R.; Hughes, D. L. *Pestic. Monit. J.* **1973,** *7,* 100–103.
36. Ohlendorf, H. M.; Risebrough, R. W.; Vermeer, K. *U.S. Fish and Wildlife Service Special Scientific Report: Wildlife No. 9;* U.S. Government Printing Office: Washington, DC, 1978.

37. O'Shea, T. J.; Brownell, R. L.; Clark, D. R.; Walker, W. A. *Pestic. Monit. J.* **1980,** *14,* 35–46.
38. Clark, D. R., Jr.; Krinitsky, A. J.; *Pestic. Monit. J.* **1980,** *14,* 7–10.
39. Hall, R. J.; Kaiser, T. E.; Robertson, W. B., Jr.; Patty, P. C. *Bull. Environ. Contam. Toxicol.* **1979,** *23,* 87–90.
40. O'Shea, T. J.; Ludke, J. L. *Monitoring Fish and Wildlife for Environmental Pollutants;* U.S. Fish and Wildlife Service. U.S. Government Printing Office: Washington, DC, 1979.
41. Cain, B. W. *Pestic. Monit. J.* **1981,** *15,* 128–134.
42. Cain, B. W.; Bunck, C. M. *Environ. Monit. Assess.* **1983,** *3,* 161–172.
43. Carson, R. *Silent Spring;* Houghton Mifflin: Boston, 1962; pp 179–180.
44. Tucker, R. K.; Crabtree, D. G. *Bureau of Sport Fisheries and Wildlife Resource Publication No. 84;* U.S. Government Printing Office: Washington, DC, 1970.
45. Heath, R. G.; Spann, J. W.; Hill, E. F.; Kreitzer, J. F. *U.S. Fish and Wildlife Service Special Scientific Report: Wildlife No. 152;* U.S. Government Printing Office: Washington, DC, 1972.
46. Hill, E. F.; Heath, R. G.; Spann, J. W.; Williams, J. D. *U.S. Fish and Wildlife Service Special Scientific Report: Wildlife No. 191;* U.S. Government Printing Office: Washington, DC, 1975.
47. Hudson, R. H.; Tucker, R. K.; Haegele, M. A. *U.S. Fish and Wildlife Service Resource Publication No. 153;* U.S. Government Printing Office: Washington, DC, 1984.
48. Hill, E. F.; Camardese, M. B. *U.S. Fish and Wildlife Service: Fish and Wildlife Technical Report 2;* U.S. Government Printing Office: Washington, DC, 1986.
49. Carson, R. *Silent Spring;* Houghton Mifflin: Boston, 1962; pp 31, 75–77, 195, 204, 235.
50. Stickel, L. F.; Chura, N. J.; Stewart, P. A.; Menzie, C. M.; Prouty, R. M.; Reichel, W. L. *Trans. N. Am. Wildl. Nat. Res. Conf.* **1966,** *31,* 190–200.
51. Ratcliffe, D. A. *Nature* **1967,** *215,* 208–210.
52. Hickey, J. J.; Anderson, D. W. *Science* **1968,** *162,* 271–273.
53. Porter, R. D.; Wiemeyer, S. N. *Science* **1969,** *165,* 199–200.
54. Heath, R. G.; Spann, J. W.; Kreitzer, J. F. *Nature* **1969,** *224,* 47–48.
55. Longcore, J. R.; Sampson, F. B.; Whittendale, T. W., Jr. *Bull. Environ. Contam. Toxicol.* **1971,** *6,* 345–350.
56. Wiemeyer, S. N.; Porter, R. D. *Nature,* **1970,** *227,* 737–738.
57. Blus, L. J.; Heath, R. G.; Gish, C. D.; Belisle, A. A.; Prouty, R. M. *Bioscience* **1971,** *21,* 1213–1215.
58. Carson, R. *Silent Spring;* Houghton Mifflin: Boston, 1962; pp 102, 185.
59. Blus, L. J. *Environ. Pollut. Ser. A* **1982,** *28,* 15–33.
60. Ludke, J. L.; Hill, E. F.; Dieter, M. P.; *Arch. Environ. Contam. Toxicol.* **1975,** *3,* 1–21.
61. Hill, E. F.; Fleming, W. J. *Environ. Toxicol. Chem.* **1982,** *1,* 27–38.

62. Grue, C. E.; Fleming, W. J.; Bushy, D. G.; Hill, E. F. *Trans. N. Am. Wildl. Nat. Res. Conf.* **1983**, *48*, 200–220.
63. Hall, R. J.; Kolbe, E. *J. Toxicol. Environ. Health* **1980**, *6*, 853–860.
64. White, D. H.; King, K. A.; Mitchell, C. A.; Hill, E. F.; Lamont, T. G. *Bull. Environ. Contam. Toxicol.* **1979**, *23*, 281–284.
65. White, D. H.; Mitchell, C. A.; Hill. E. F.; *Bull. Environ. Contam. Toxicol.* **1983**, *31*, 93–97.
66. Rattner, B. A.; Sileo, L.; Scanes, G. C. *Pestic. Biochem. Physiol.* **1982**, *18*, 132–138.
67. Eastin, W. C., Jr.; Fleming, W. J.; Murray, H. C. *Comp. Biochem. Physiol.* **1982**, *73C*, 101–107.
68. Haseltine, S. D.; Hensler, G. *Environ. Pollut. Ser. A* **1982**, *25*, 139–147.
69. Hoffman, D. J.; Eastin, W. C., Jr.; *Environ. Res.* **1981**, *26*, 472–485.
70. Carson, R. *Silent Spring*; Houghton Mifflin: Boston, 1962; pp 31, 75–77, 195, 235, 294.
71. Fleming, W. J.; Clark, D. R., Jr.; Henny, C. J. *Trans. N. Am. Wildl. Nat. Res. Conf.* **1983**, *48*, 186–199.
72. Blus, L. J.; Henny, C. J.; Kaiser, T. E.; Grove, R. A. *Trans. N. Am. Wildl. Nat. Res. Conf.* **1983**, *48*, 159–174.

7 ∾ Human Health Effects of Pesticides

J. E. Davies and R. Doon

Rachel Carson was the first to challenge modern society with the need for an assessment of the health effects of pesticides. She realized that pesticides posed health threats because they were highly toxic and also persistent and ubiquitous in our environment. *Silent Spring* expressed concerns for the possible acute and chronic health sequelae after exposure to these xenobiotic agents.

In the ensuing years, health effects reflective of both acute and chronic toxicity, especially due to the persistent pesticides, have been noted in all parts of the world. The type of health effects noted when exposures were excessive were related to the specific pesticide being studied; different diseases were observed in different exposure situations.

Careless application frequently was the mechanism for acute poisonings, whereas the occupationally exposed worker might have a more sustained type of exposure producing different types of diseases. Examples of the types of illnesses encountered include dermatitis, kidney problems, and adverse reproductive effects.

Whether a disease developed or not was dependent on the exposure dose received by the worker. Besides acute poisoning and chronic occupational exposure, a third type of exposure, called incidental pesticide exposure, is the type of exposure received by the general public because of the presence of trace

0980-4/87/0113$06.00/0 © 1987 American Chemical Society

amounts of pesticides in the daily environment. These traces can be detected in human fat, urine, or mother's milk, and the compounds most frequently identified are the chlorinated hydrocarbon pesticides. Their presence in the general public has been one of the factors that prompted regulation and even cancellation of these more persistent pesticides.

When public health investigators have conducted research with these different types of exposures, four pesticide management problems have been recognized: (1) poisonings and other health effects, (2) persistence, (3) resistance, and (4) disposal. The priorities of these problems have changed with time and vary enormously in severity and magnitude in the developing world as compared to the developed world.

These pesticide management problems are of equal concern to public health and agriculture authorities, and each will require carefully thought-out strategies if they are to be resolved. The Consortium for International Crop Protection (CICP), which is a consortium of U.S. universities concerned with crop protection, has recognized the contribution of both health and agriculture to pest management. CICP has great concern for the effectiveness of global pesticide management. Because of the four pesticide management problems, CICP recommended an agromedical approach. The term *agromedicine* has been defined as the integrated interdisciplinary application of the skills and knowledge of agriculture, applied chemistry, and medicine to the safe global production of enough food of high nutritious content to meet the health and nutritional needs of humans. The potential of agromedicine has applicability to the four major pesticide management problems.

Acute Pesticide Poisoning

In spite of the obvious public health importance and interest, the exact number of pesticide poisonings is not known. Different strategies have been used to ascertain the magnitude of the problem. The Director General of the World Health Organization (WHO) sent a questionnaire to all member nations asking them what was the status of pesticide poisoning in their countries. Replies were received only from 10 countries.

The data were used to make a model on pesticide poisoning incidence. WHO calculated from this model that 500,000 pesticide poisonings per year occurred with a 1% case fatality rate (1).

Copplestone (2) estimated, on the basis of the world population being 4000 million, the actual number of pesticide poisoning deaths might be 20,640 per year. Because these were statistics of accidental poisonings, he argued, "they should be totally preventable providing that the technology of safety was transmitted to the exposed worker in the developing countries."

Suicide is another mechanism of pesticide poisoning and death, and it is clearly on the increase in the developing world. Surveys have emphasized a need for national programs to limit the ready availability of highly toxic chemicals such as pesticides in the home setting. Suicide with pesticides is reaching almost epidemic proportions in certain areas of the world. More than 1000 deaths due to pesticide poisoning occurred in Sri Lanka in 1978 (3). Suicide attempts accounted for 73% of the total. During the same year only 572 deaths were due to polio, diphtheria, tetanus, and whooping cough, and there were no deaths due to malaria. During a 3-year period in the Philippines, 60.17% of pesticide poisoning cases were due to suicide (4).

Suicide has also shown a dramatic increase in Trinidad-Tobago. A sharp rise occurred in 1965, shortly after the importation of highly toxic pesticides into the country. A survey of chemical intoxication cases admitted to the three hospitals on the islands showed that 83% of the deaths resulted from pesticide poisoning; 77% of the 93 occurring in 1984 were due to paraquat; 69% of poisonings occurred in East Indians (R. Doon, unpublished data).

Copplestone (2) concluded, "In all countries, but especially where literacy is low, probably the most effective single measure to promote safety is restriction of the availability of the more toxic pesticides to those who have been trained in their use and who have a specific need to use them." To assist countries in furthering this type of control, WHO has adopted a recommended classification of pesticides by hazard (5). This decision has been a major step toward classification of pesticides in developing countries where poisoning appears to be out of control.

Other Health Effects

Numerous epidemiologic studies have been conducted in the occupationally exposed, and adverse health effects have been noted chiefly during the manufacture of single groups of pesticides. Neurological and neurobehavioral diseases were also observed during the formulation and excessive exposure sustained during the manufacture of chlordecone (Kepone) in Hopewell, Virginia (6). Many workers demonstrated opsoclonus (nonrhythmic muscle spasms of the eyes), which was correlated with blood chlordecone levels.

Another example occurred during the manufacture and formulation of leptophos (7). Leptophos had earlier caused hind-quarter paralysis in buffalo in Egypt (8). At a later date chronic neurologic disease resembling multiple sclerosis was noted in workers who were manufacturing the compound in the Velsicol plant in Houston, Texas (9).

Reproductive adverse health effects have been found during the application of dibromochloropropane (10). Male sterility was recognized as a result of occupational exposure to this fumigant. This adverse health effect resulted in its cancellation in the United States; subsequent follow-up of the exposed workers has suggested that the testicular effects were reversible.

With regard to DDT exposure, apart from selective changes in the liver enzymes that metabolize drugs and hormones, and increases in fats and triglycerides in plasma, adverse health effects from exposure have not been unequivocally demonstrated. In those who received larger doses of DDT exposure because they were manufacturing the compound, a 57% increase in excretion of urinary hydrocortisol was demonstrated in 18 persons with a 5-year history of work exposure in a DDT manufacturing plant. Serum phenylbutazone half-life was reduced by 19% when compared to 18 matched controls (11). This means that people with high DDT exposure, because of the enzyme induction in the liver, were able to metabolize the compound phenylbutazone quicker than those without this DDT exposure. These studies suggested that there might be subtle changes taking place in the body's ability to metabolize drugs and hormones.

The interaction of drugs and pesticides in the body was first demonstrated by Kolmodin et al. (*12*), who showed increased metabolism of antipyrene under conditions of occupational exposure to a variety of organochlorine pesticides (lindane, DDT, and chlordane). The implications of drug and pesticide interactions have received scant attention, and the pharmacological consequences of these interactions is worthy of further study.

Cancer Risks

Cancer is of universal concern and interest. Many of the pesticides that have been widely used in the past have been shown to be carcinogenic on the basis of animal testing or short-term in vitro mutagenicity testing. In the United States, several pesticides have either been canceled or severely restricted because of an oncogenic trigger. In a positive epidemiologic study, Mabuchi et al. (*13*) demonstrated an association between occupational exposure to inorganic arsenic and lung cancer. This evidence was particularly impressive because it suggested a dose effect. The relationship of pesticides to cancer is the subject of continuing research, controversy, and concern. Toxicologic data rather than epidemiologic has usually provided the evidence for the cancer trigger.

Extrapolating cancer risks from mice or rodents to humans is difficult, and good epidemiologic data are more convincing because the risk in humans is measured. However, epidemiologic studies of pesticides for cancer risks are complicated by shortcomings in human exposure data, the multiplicity of pesticide exposures, changes in pesticide-use patterns, a rapid turnover of employees, and the latency of cancer.

In spite of these difficulties, however, in the last decade a sizable body of epidemiologic data has appeared in the medical literature and is beginning to contribute to a better understanding of the carcinogenic risk to humans. With regard to cancer, with the exception of inorganic arsenic (for which occupational exposure has been linked to the subsequent development of lung cancer), the data from epidemiologic studies have on the whole been reassuring because pesticide-linked carcinogenesis has not been widely proven except as just mentioned.

Persistence

Persistence is the second problem and is due to the lipophilic potential of certain types of chemicals. *Lipophilic* chemicals are those that are able to dissolve in fat and not in water. These pesticides are stored in the fat and in the food, and become residues in food and human tissues.

Residues in Food

In food, residues have had a serious economic impact. For example, in El Salvador where the meat export trade is worth more than $1 million, repeated testing showed unacceptable DDT residues. These residues were due to the use of DDT for agriculture and the subsequent grazing of the cattle in DDT-treated cotton fields where they had wandered from the pasture lands. In addition, cattle in pens were fattened with cottonseed oil that contained high residues of DDT. The result was unacceptable residue values and a net loss of more than $1 million nearly every year. The same problem is surfacing increasingly in different countries as residue analyses of imported food have greater acceptance.

Residues in Humans

As has already been mentioned, trace amounts of pesticides occur in the food, air, and water; they can be absorbed through the intact skin, inhaled, or ingested in the food and water. The chlorinated hydrocarbons are stored in body fat or excreted in urine or mother's milk. Food is an important source of these exposures, but for DDT, house dust is also an important mechanism of contamination (14, 15).

In the years since *Silent Spring*, as analytical methods have developed, the list of pesticides that have been recognized as part of the human body burden has grown.

DDT was the first compound to be recognized; later other

organochlorine pesticides were identified. All forms of life have been shown to contain trace amounts of DDT. They have been detected in almost every tissue of the human body (*16*). Similarly, alkylphosphate and phenolic metabolites, which are derivatives of certain carbamates and organophosphate insecticides, are readily detected in the urine of the average U.S. citizen. Pesticide residues have not provided evidence of disease, being rather a measure of the small exposures that are ubiquitous in our modern environment. Nevertheless, their recognition in the U.S. general population contributed toward their ultimate cancellation.

In the United States, pesticide residue monitoring studies have been conducted on an annual basis, and in 1975 the following pesticides were identified: DDT, hexachlorobenzene, α-BHC, β-BHC, γ-BHC, dieldrin, oxychlordane, and *trans*-nonachlor (*17*).

In the United States, these surveys were based on a stratified sample of the population, and as can be seen from Table 1, residues of DDT, dieldrin, and β-BHC have shown a healthy decrease with time. DDT was cancelled in 1972, but the decline had already started before the cancellation. DDT residues in the black population were twice as high as in the whites (*18*).

Global monitoring of organochlorine pesticides has been an area of special concern to the World Health Organization and the United Nation's Environmental Program. These two agen-

Table 1. Changes over Time of the Geometric Means of Selected Organochlorine Residues in the Adipose Tissue of the U.S. Black and White Populations from 1970 through 1978

Year	Persons	Total DDT		Dieldrin		β-BHC	
		Black	White	Black	White	Black	White
1970	1373	12.97	7.46	0.35	0.22	0.69	0.44
1971	1545	16.21	7.04	0.35	0.24	0.65	0.37
1972	1877	12.16	6.23	0.27	0.21	0.47	0.30
1973	1085	9.66	5.70	0.27	0.21	0.37	0.32
1974	892	8.44	4.96	0.19	0.18	0.31	0.23
1975	771	7.04	4.67	0.16	0.16	0.27	0.24
1976	667	6.70	4.38	0.16	0.14	0.30	0.25
1977	775	4.47	3.35	0.11	0.12	0.19	0.20
1978	802	6.07	3.64	0.14	0.12	0.23	0.19

cies, through their Global Environmental Monitoring System (GEMS), have studied human exposures to selected organochlorine pesticides (19), monitoring mother's milk collected from the following countries: Belgium, India, Israel, Mexico, Federal Republic of Germany, Japan, People's Republic of China, Sweden, and Yugoslavia.

The median levels of DDT in China, India, and Mexico were severalfold higher than those from other participating countries. The high levels in China, India, and Mexico most certainly reflected the continued use of DDT in agriculture and for malaria control.

We, too, have studied pesticide levels in some of the countries in the Caribbean. We conducted limited serological surveys of organochlorine residues throughout the Caribbean and, here too, we found noticeable differences. The highest levels of DDT were found from Haiti, a finding that was not surprising because DDT is still used for malaria control in this country.

Regarding dieldrin, we were surprised to find that more than 80% of serum samples from one of the islands were positive for trace amounts of dieldrin, compared to only 9% in the United States. A field investigation demonstrated that dieldrin was still being used on vegetables for control of mole crickets, cut worms, and ants (20). Dieldrin was applied by pouring the concentrate into a water-filled ditch and scooping the solution up in a bucket and throwing it onto the adjoining rows of vegetables. Almost certainly this agricultural use of dieldrin was the explanation for these high residues noted in the general population (R. Doon, unpublished data).

Studies such as these emphasize the agromedical potential of human monitoring studies. The public health monitoring of the general population facilitated the identification of undesirable agricultural use of a pesticide that was contaminating the general public. The situation has now been cleared up, and the general body burden profile more closely resembles that seen in the other nations of the region.

Resistance

In 1948, a resistance of the housefly to DDT first became apparent in Italy, Sardinia, and the United States (21). By 1949,

DDT-resistant mosquitoes had been found in the United States. From these small beginnings, the resistance problem increased by leaps and bounds. Brown (22) reported that since 1945 the total number of species with resistant strains had risen to 137, about equally divided between pests of agriculture and medical–veterinary importance. By 1965, the results of a world survey of resistance of agricultural pests including insects, mites, plant pathogens, and rodents indicated that strains of some 200 species had become resistant (23). By 1980, resistant strains of some 414 arthropod pests, 152 of which were medically important, had been noted (24).

Dover and Croft (25) advocated the philosophy of "resistance management", which should be considered part of an overall integrated pest management program. This approach is furthered by appropriate pesticide regulation, incorporating resistance risk into pesticide registration, and allowing mixtures to be registered for use in resistance management.

Disposal

This is the fourth and most pressing pesticide management problem in the United States. It first surfaced with the Love Canal episode and subsequently led to the special assignment of "Superfund" monies to help clean up toxic wastes in the United States. As environmental monitoring has increased, so has the magnitude of the hazardous waste sites become apparent; there is currently special concern for the long-term impact of ground water contamination.

For example, in Florida, aldicarb and ethylene dibromide are two very different types of pesticides that have caused public health concerns because they were identified in ground water systems of the state.

Although the disposal problem has not yet been addressed, it is probably one of the most pressing problems in the Third World. Preliminary surveys in Indonesia and Southeast Africa show that sizable collections of unwanted deteriorating supplies of pesticides are scattered throughout the countryside. Some of these reflect exports of donated materials from donor countries to Third World countries where not all of the materials are used up; now, as they begin to break down, they pose a serious threat to rural communities (26). Pesticide disposal, therefore, is a demanding and vitally important problem for the developed and developing world.

The Years Ahead

Silent Spring was the spur that prompted a plethora of research to ascertain potential and real health effects of technology. In the years between, significant progress has been made in regulation of pesticides and protection of citizens in the developed world. Widespread pesticide poisonings do not currently appear to be a major problem in the United States. However, we must not become complacent; further research should be conducted to determine the safety of present agricultural chemicals to ensure the good health of future generations.

In the meantime, the lessons learned from various epidemiologic studies are clear, and the marriage of integrated pest management (IPM) and agromedicine must be the prescriptio1 for pest management in the years ahead. The total picture is still not generally known with regard to the health effects of the multiplicity of pesticides in use. In these circumstances no one will question the fact that any excessive exposure is unwarranted, and good agromedical practices should strive to minimize exposure.

Regulations, training, and education have furthered the general health of pesticide workers and in certain areas have curtailed their exposure. Cancellation orders have removed some of the more persistent pesticides and have resulted in a demonstrable reduction of general population body burdens. The ultimate solution to acute poisoning rests in the need for action through control, plus continued education and training programs being extended to the developing world.

Insecticide resistance as well as vector resistance problems are only too apparent in the developing world. The needs, therefore, for the problem of resistance are integrated approaches for pests and vectors.

Food and human residues are monitored very little, so data gaps exist for information on human and environmental pesticide pollution. The resolution of these problems will call for the agromedical approach, and a successful outcome will require agromedical training at the small farmer level, the development of appropriate regulations and importation policies at the national level, and guidelines for the technology in the already experienced developed world. A judicious importation policy that avoids the more toxic chemicals getting into the hands of the inexperienced rural farmer will be needed, together with the exclusion of the persistent pesticides and the resultant pesticide residue problems.

In addition, pesticide management policies should respect the requirements of the *Codex Alimentarius,* the Guide to Codex Maximum Limits for Pesticide Residues from the Food and Agriculture Organization (FAO), Rome.

I believe these recommendations are some of the ways in which future generations will be protected.

Literature Cited

1. *Safe Use of Pesticides.* 20th Report, World Health Organization to the Expert Committee on Pesticides, WHO Technical Report, 1973. Series 513, pp 42–43.
2. Copplestone, J. F. In *Global View of Pesticides;* Watson, D.; Brown, A. W. A., Eds.; Academic: New York, 1977; pp 147–157.
3. Jeyaratnam, J.; de Alwis Seneviratne, R. S.; Copplestone, J. F. *Bull. WHO* **1982,** *60(4),* 615–619.
4. Zapatos, C. C., M.S.P.H. Thesis, University of Miami School of Medicine, June 1982.
5. *WHO Chron.* Recommended Classification of Pesticides by Hazard; **1975,** *29,* 397–401.
6. Cohn, W. J.; Boyland, J. T.; Blanke, R. V. Fariss, M. W.; Howell, J. R.; Guzelian, P. S. *N. Engl. J. Med.* **1978,** *298,* 243–248.
7. American Occupational Medicine Association (AOMA) Report, January 1978.
8. Near East News Roundup, Food Agriculture Organization, RNEA, Cairo, November 22, 1971.
9. Xintaras, C.; Burg, J. R.; Tanaka, S.; Lee, S. T.; Johnson, B. L.; Cottrill, C. A.; Bender, J. *Occupational Exposure to Leptophos and Other Chemicals,* DHEW (NIOSH) Publication No. 78-136, U.S. Department of Health, Education, and Welfare: Cincinnati, OH, 1978.
10. Whorton, D.; Krauss, R. M.; Marshall, S.; Milby, T. H. *Lancet* **1977,** *2,* 8051.

11. Kolmodin-Hedmin, B. *Eur. J. Clin. Pharmacol.* **1973,** *5,* 195–198.
12. Kolmodin, B.; Azarnoff, D.; Sjoquvist, F. *Clin. Pharmacol. Ther.* **1969,** *10,* 638.
13. Mabuchi, K.; Lillienfeld, A. M.; Snell, L. M. *Prev. Med.* **1980,** *9,* 51–77.
14. Mrak, E. M. *Report of the Secretary's Commission on Pesticides and Their Relationship to Environmental Health;* U.S. Government Printing Office: Washington, DC, 1969; Parts 1–2.
15. Davies, J. E.; Edmundson, W. F.; Raffonelli, A. *Am. J. Public Health* **1975,** *65(1),* 53–57.
16. *An Agromedical Approach to Pesticide Management: Some Environmental Considerations;* Davies, J. E.; Freed, V. H.; Whittemore, F. W., Eds.; University of Miami Press: Miami, 1983.
17. Kutz, F. W.; Yobs, A. R.; Strassman, S. C.; Viar, J. R., Jr. *Pestic. Monit. J.* **1977,** *11(2),* 61–63.
18. Davies, J. E.; Edmundson, W. F.; Schneider, N. J.; Cassady, J. C. *Pestic. Monit. J.* **1968,** *2(2).*
19. Global Environmental Monitoring System, *Report: Assessment of Human Monitoring;* Slorach, S. A.; Vaz, R., Eds.; Swedish National Food Administration: Uppsala, Sweden, 1983.
20. Poland, A.; Smith, D.; Kuntzman, R.; Jacobson, M.; Conney, A. H. *Clin. Pharmacol. Ther.* **1970,** *11,* 724.
21. Patton, T. E.; Whittemore, F. W. *New Jersey Mosquito Extermination Association Proceedings,* 1950.
22. Brown, A. W. A. *Bull. Entomol. Soc. Am.* **1961,** *7,* 6–19.
23. *Report of the First Session of the FAO Working Party of Experts on Resistance of Pests to Pesticides;* Food Agriculture Association (FAO): Rome, 1967.
24. Georghious, G. P. *Residue Rev.* **1980,** *76,* 131–145.
25. Dover, M.; Croft, B. *Getting Tough: Public Policy and the Management of Pesticide Resistance;* World Resources Institute: Washington, DC, 1984; Study No. 1.
26. Jensen, Janice *Pesticide Use in East and Southern Africa—An Overview;* proceedings from a regional workshop, Nairobi, Africa, March 1985, pp 48–51.

Overviews

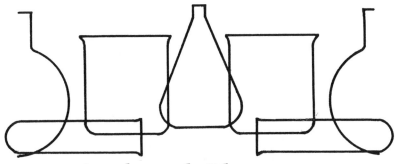

8 ∽ Analytical Chemistry of Pesticides: Evolution and Impact

Joseph D. Rosen and Fred M. Gretch

Rachel Carson's *Silent Spring* alerted the public to the potential environmental dangers of indiscriminate pesticide use as well as the lack of testing of pesticides for long-term effects on human health. This public awareness and concern was soon translated into a flow of research dollars for study of the fate of pesticides in the environment; their concentrations in the biosphere; and their metabolism, pharmacokinetics, and toxicology. It soon became apparent that sensitive analytical methods were needed to conduct these studies. Within a few years, detection levels were reduced from the parts-per-million (ppm) level to the parts-per-billion (ppb) level. Further advances in analytical methodology in the past few years have brought the limit of detection down to the parts-per-trillion (ppt) level. These advances have not been matched by advances in risk assessment and risk management, mainly because of the inability to determine cancer incidence experimentally at low exposure levels as well as the inability to establish a threshold value for carcinogens. Thus we are in the unhappy position of being unable to assess the risk of chemicals whose quantities we can easily measure.

0980–4/87/0127$06.00/0 © 1987 American Chemical Society

How Small Is Small?

To appreciate how small these concentrations are and how truly remarkable the achievements of the analytical chemist are, consider that 1 ppm is the equivalent of 1 inch in 16 miles, a drop of vermouth in 1000 quarts of gin, or approximately one mouthful of food in a lifetime. One part per billion represents a concentration 1000 times less than 1 ppm and is the equivalent of 1 second in 32 years or one pinch of salt in 2000 1-lb bags of potato chips. An analytical detection limit of 12 ppt literally represents the proverbial "needle in a haystack", provided that the haystack weighs 100,000 tons.

The Analytical Chemist's Tasks

In addition to finding the "needle" in so much "hay", the analytical chemist had to devise methods to quantify many different needles (pesticides) in many different haystacks (numerous foods, soil, water, and various animal tissues).

A bigger problem than most people realize was the quantitative extraction of residues from samples; that is, removing (extracting) the needle from the haystack and determining the exact weight or quantity removed. Furthermore, analysts had to find techniques for eliminating or at least significantly reducing the quantity of interfering substances; in other words, to clean up the crude sample extracts. Finally, detection methods were needed to be both selective and sensitive for the levels of residues anticipated. These challenges were met with an amazing degree of success, considering the difficulties involved. Almost every available physical, chemical, and biological technique was evaluated in the search for workable methods. As new analytical "tools" were developed in the research laboratories, they were quickly added to the arsenal of the residue analyst.

Evolution of Analytical Methodology

Colorimetric Determinations

Colorimetric determinations are those in which the substance being analyzed is made to form a color. The amount of light absorbed by the colored compound is then compared to the amount of light absorbed by known amounts of this material. Most of the routine methods in use at the time of publication of *Silent Spring* relied on colorimetric determinations. The classic method of Schechter et al. for DDT (1) involved the extraction of the residue from the sample followed by a chemical reaction to produce a specific and reasonably stable color. The intensity of the color was measured spectrophotometrically, that is, by determining the amount of visible light that the solution absorbs. Then, the quantity of the residue was calculated from a calibration curve prepared from solutions of known concentrations of DDT that had been treated similarly. Similar procedures were used to determine diazinon (2), heptachlor (3), and dieldrin (4). Many of these colorimetric methods were sensitive at the parts-per-million level and below but suffered from numerous disadvantages: lengthy analysis time, the need to analyze large quantities of sample, inability to analyze for more than one pesticide at a time, and a lack of specificity.

Sample Cleanup

The determination of pesticide residues requires analytical methods of extreme sensitivity and precision which, for the most part, are reliable only in the absence of significant quantities of interfering substances. In the Schechter–Haller method for DDT, cleanup of the sample extract was achieved by chemical oxidation of extraneous material. Advances in sample cleanup were achieved by judicious use of organic solvent mixtures (solvent partitioning) and absorption of interfering materials on magnesium silicate columns.

Jones and Riddick (5) discovered that some synthetic organic pesticides could be separated from interfering biological materials by partitioning between acetonitrile and *n*-hexane. This procedure involved repeated acetonitrile partitioning of the sample extract dissolved in *n*-hexane and then evaporating the combined acetonitrile phases just to dryness. A large proportion of the fats and waxes would thus remain in the hexane phase. The residues were subsequently dissolved in an appropriate solvent for colorimetric determination of the specific pesticide.

Ordas et al. (3) suggested the use of a synthetic magnesium

silicate (Florisil) as an adsorbent for the removal of lipids and pigments from sample extracts. They demonstrated that both heptachlor and chlordane could be eluted from a column of activated magnesium silicate with solvent and that this procedure provided an adequate cleanup for the colorimetric determination of these two pesticides in a wide variety of food and forage crops.

Bioassay Methods

An interesting approach to the measurement of the relative quantity of residue in foods was the development and adaptation of bioassay methods. The major disadvantage was a lack of specificity, but at the time of their development they were capable of measuring quantities of toxic residues with a sensitivity unattainable by most chemical methods. Frawley et al. (6) used a procedure for the bioassay of insecticides by oral administration to houseflies. An ether extract of a food sample was evaporated to dryness over sugar, and an aqueous solution of the residue was then fed to houseflies previously fasted for 24 hours. The method was sensitive to 3 μg of parathion, 4 μg of EPN, 5 μg of lindane, and 15 μg of DDT. (A *microgram* (μg) is one millionth of a gram.) For samples of unknown spray history, however, bioassay methods were useless.

Simultaneous Detection of Multiple Pesticides

A *tolerance* is the amount of a particular pesticide residue allowed by the government on a specific food. The amounts can vary for the pesticide and the food. Tolerance exemptions were allowed for pesticides thought to have little mammalian toxicity.

By 1962 more than 2000 tolerances or exemptions from tolerances had been approved by the Food and Drug Administration. This number reflects the use of more than 100 pesticide chemicals on specific raw agricultural commodities. The residue analysts concerned with enforcing those tolerances could no longer afford the luxury of using individual methods to test for each possible residue. Said Paul Mills of the Food and Drug Administration, "If a regulatory chemist is given a perishable sample of unknown spray history (say a fruit or fresh vegetable) to check for all possible spray residues, it might well take him a month to apply all the known specific techniques. By that time, his sample would have deteriorated, and the shipment of the fruit or vegetable would have gone into consumption long since." Mills (7) published the first systematic approach to the determination of multiple pesticide residues in foods. Solvent extraction procedures were provided for both nonfatty (fruit and vegetables) and fatty (fats, oils, cheese, milk, and animal tissue) foods. By using paper chromatography (8) for separation and detection, and both solvent partitioning and magnesium silicate column chromatography for sample cleanup, it was possible to identify and semiquantify parts-per-million residues of DDT, TDE, DDE, BHC, lindane, methoxychlor, toxaphene, perthane, chlordane, heptachlor, heptachlor epoxide, dieldrin, endrin, aldrin, and kelthane.

Gas Chromatography (GC)

Paper chromatography is an analytical procedure to separate and quantify components of a mixture. The mixture is placed on an absorbent paper, and the components of the mixture are separated with the aid of a solvent. Paper chromatography is tedious, and it is much inferior to gas chromatography in terms of speed, convenience, resolution, and overall utility. *Gas chromatographic* analytical procedures separate chemicals by taking advantage of their boiling point and solubility differences. The sample to be analyzed is vaporized, separated, and then detected electrochemically. The length of time a substance remains on the column is called the *retention time*. The substance being analyzed is then *eluted*, that is, removed from the column by a stream of gas. When two substances are removed together, they *coelute*.

Originally, the use of GC was limited to analysis of pesticides in formulations because GC detectors in the early 1960s were either relatively insensitive (the thermal conductivity detector) or sensitive but not selective (the flame-ionization detector). Widespread use of GC for pesticide residue analysis was made possible by advances in detector technology.

The development by Coulson et al. (9) of a halogen-specific

microcoulometric detection system for GC allowed for the analysis of halogenated pesticides, which were then very common in environmental samples. In *microcoulometric* analysis, the amount of substance present is determined by measuring the amount of electricity required to remove the substance from a solution. This discovery occurred at an opportune time, because the controversy generated by *Silent Spring* soon resulted in research money that facilitated the development and commercialization of this detector and the others that followed. Burke and Johnson (10) evaluated the microcoulometer and successfully chromatographed 71 pesticides. The possibility of performing microcoulometric analyses of crop samples without prior cleanup was also investigated, with unsatisfactory results. Recoveries in general were low, and in some instances, heptachlor was not recovered at all. The chromatography of DDT and endrin was affected; the typical chromatographic peaks were rearranged or shifted. The crop material contaminated the chromatographic column to such an extent that it was unsuitable for further use. It was concluded that it was necessary to clean up all blended extracts to eliminate interferences prior to GC analysis (10).

Mills et al. (11) combined and modified portions of previously reported cleanup procedures to provide a rapid, simple, and routinely applicable method for quantitative analysis of residues of organochlorine pesticides in nonfatty foods. Pesticide residues were extracted with acetonitrile. The filtered acetonitrile extract was diluted with water and partitioned with petroleum ether; the petroleum ether extract was further cleaned up by magnesium silicate column chromatography. Pesticides were identified and quantified by microcoulometric GC. Good recoveries were obtained for five pesticides added to 11 products (fruits, vegetables, and canned soups) at levels of 0.02 to 0.2 ppm (i.e., 20 to 200 ppb).

The development of another GC detector (electron-capture detector) (12) had a profound effect on the field of pesticide residue analysis. The electron-capture detector was much more sensitive to very small quantities of chlorinated organic compounds than the microcoulometric detector and made it possible to determine, quickly and accurately, levels of chlorinated pesticides that could not have been measured previously.

Although the electron-capture detector was extremely sensitive to chlorinated compounds, it was not entirely specific. This fact was pointed out by Burke and Giuffrida (13), who stressed the need for adequate cleanup before a sample was injected into the electron-capture gas chromatograph. Poorly cleaned-up extracts were shown to contaminate a column and result in weak or false detector responses. Satisfactory results for a wide variety of nonfatty foods were obtained by combining the Mills et al. (11) extraction and cleanup procedure with electron-capture GC under a set of standardized operating conditions. Analysis of broccoli, fortified with seven common pesticides (lindane, heptachlor, heptachlor epoxide, aldrin, dieldrin, endrin, and DDT) at levels from 0.001 to 0.1 ppm, gave recoveries that varied from 73% to 98%.

In 1965, Coulson (14) introduced an electrolytic conductivity detector for GC. This detector was highly selective for halogen, sulfur, and nitrogen atoms. Then, Hall (15) developed a micro-electrolytic conductivity detector that was 20 to 50 times more sensitive than the Coulson detector. The currently available commercial version (the Model 700A Hall electrolytic conductivity detector) is about 2 to 5 times less sensitive than the electron-capture detector but is much more specific. It is about 1000 times more sensitive than the microcoulometric detector and much easier to operate and maintain.

Specific detectors for pesticides containing phosphorus, sulfur, nitrogen, or all three were also developed during this period.

Giuffrida (16) modified a conventional gas chromatographic flame-ionization detector by fusing an alkali salt onto the electrode. This thermionic detector exhibited responses to organophosphorus pesticides up to 600 times that achieved with a conventional flame-ionization detector. Organophosphorus pesticides in vegetables were easily detected by the thermionic detector at levels of 0.05 and 0.1 ppm after sample cleanup by the method of Mills et al. (11). Crop materials did not interfere even when the equivalent of 5 g of original sample was injected.

Brody and Chaney (17) developed a flame photometric detector for the determination of sulfur- and phosphorus-containing compounds. This detector has found increasing use in residue analysis of organophosphate and organothiophosphate pesticides. It is highly sensitive and selective and is considered an equal or better alternative to the thermionic detector for the determination of organophosphorus pesticide residues.

Kolb and Bischoff (18) described the first thermionic detector sensitive to both nitrogen- and phosphorus-containing pesticides. This detector (named the N/P detector) is highly selective, stable, sensitive, and reproducible in its response to phosphorus or nitrogen.

As detection limits were being reduced by advances in GC and cleanup techniques, it soon became apparent that, in many cases, other components of the material that was being analyzed or artifacts introduced by the analytical procedure were giving detector responses at retention times identical to many pesticides. For example, the coelution of polychlorinated biphenyls (PCBs) with DDT and DDE (Figure 1) resulted in either misidentification of the latter two materials or highly inflated concentrations for them in environmental samples that were PCB-contaminated (19). In recognition of such problems, analysts consider retention time data on one chromatographic column as tentative evidence. Confirmation of residue identity is made by reinjecting the sample on a chromatographic column of different polarity. Table 1 lists the retention times of several pesticides on two chromatographic columns of different polarity. Dieldrin and DDE have identical retention times on Column A (OV-101) but different retention times on Column B (mixed OV-101 and QF-1). Laboratories doing pesticide residue analyses have an extensive list of pesticide relative retention times on two different columns; this list is used for more reliable identification. In most cases, this system works well, but sometimes this system can cause misidentification, as in the case of a sample of onions from the Netherlands that was thought to contain the organophosphorus pesticide bromophos on the basis of retention time data obtained on two different chromatographic columns and four different detection systems (electron capture, microcoulometry, flame photometry, and nitrogen–phosphorus). Analysis by a more specific detection method, gas chromatography–mass spectrometry (GC–MS), indicated that the residue was a different organophosphorus insecticide, trichloronate (20).

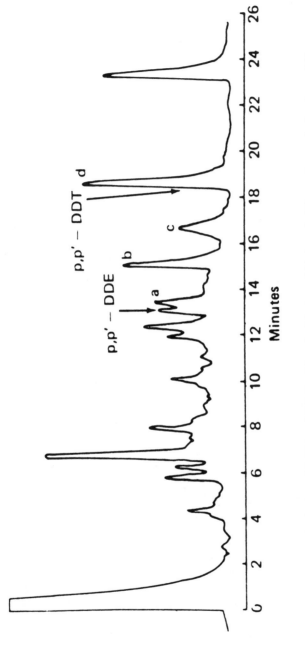

Figure 1.   Coelution of DDT and DDE with polychlorinated biphenyls. The peaks marked a–d are different polychlorinated biphenyls. (Reproduced with permission from reference 19. Copyright 1971 Springer Verlag.)

Table 1. Relative Retention Times of Several Pesticides
on Different GC Columns

Pesticide	Column A^a	Column B^b
Dieldrin	1.83	2.22
DDE	1.81	1.88
Parathion	0.96	1.88
Malathion	0.87	1.48
Linuron	0.83	1.50

NOTE: Relative to aldrin = 1.

^{a}OV-101.
^{b}OV-101–QF-1.

Gas Chromatography–Mass Spectrometry (GC–MS)

A *mass spectrometer* is an instrument that is capable of fragmenting a chemical compound into its constituent parts (*ions*), separating these ions, and recording each ion separately. Such a recording is known as a *mass spectrum*. The mass spectrum is characteristic of the chemical and may therefore be used to definitively identify the chemical, much the same way as fingerprints identify people. Mass spectrometers require very little sample and operate very rapidly. They are thus very useful in identifying minute amounts of chemicals in a mixture, provided that these chemicals can be separated from each other, a task easily accomplished by a gas chromatograph.

The marriage of GC, an excellent separation technique, to MS, an excellent identification technique, has proved to be one of the most outstanding advances in pesticide residue analysis because it can provide full mass spectra of pesticides at the nanogram (10^{-9} g, billionth of a gram) level in complex mixtures. Pesticide residue confirmation at subnanogram levels can be obtained if the analyst uses selected-ion monitoring (SIM), a technique that turns the mass spectrometer into an even more sensitive and very selective detector for gas chromatograph eluates (21). With SIM, the mass spectrometer can be instructed to detect either one or several of the ions generated in its ion source, and therefore any contaminant can be detected and quantified provided that the contaminant can be chromatographed and its mass spectrum is known. Thus, instead of using the selectivity features for halogens as in electron-capture detection or phosphorus atoms as in thermionic detection, the known fragment ions of the pesticide produced by the mass spectrometer become the detectable species.

Medium-resolution selected-ion monitoring provides for even greater sensitivity because of enhanced selectivity. This technique allows for the detection of exact mass rather than nominal mass ions. For example, analysis of the potent carcinogen aflatoxin B_1 (at mass 312) in peanuts is masked by the presence of an ion of the same nominal mass that comes from peanut oil. The exact mass of the aflatoxin B_1 ion is 312.0633, whereas that of the peanut oil fragment is 312.2183. These two ions can be separated by operating the mass spectrometer at medium resolution, thus allowing for the analysis of aflatoxin B_1 in peanuts at the 100-parts-per-trillion level (22). This result is a clear example of the increases in sensitivity attainable if selectivity is enhanced.

Concurrent with advances in selective detection have been advances in GC column technology. Capillary column GC gives better separations, leading to higher sensitivity, higher precision, more rapid analysis, and the ability to chromatograph polar and high boiling materials. Fused silica capillary columns now make possible the analysis of numerous materials heretofore very difficult or impossible to analyze by GC. Advances in high-pressure liquid chromatography and coupling of liquid chromatographs to mass spectrometers by either moving belt or thermospray interfaces will offer additional choices for sensitive and confirmatory methods for the analysis of carbamate and triazine pesticides as well as polar metabolites and environmental degradation products.

Two other advances in MS deserve mention. Negative-ion chemical ionization (NICI) is yet another technique that allows greater selectivity and sensitivity in analysis. Materials that contain atoms capable of capturing electrons (most pesticides fit into this category) are particularly easy to analyze by this technique. Stout and Steller (23), for example, were able to detect as little as 5 picograms (5 trillionths of a gram) of pendimethalin, the active ingredient in PROWL herbicide, using NICI. Today, tandem mass spectrometry (MS–MS) opens up a whole new dimension in selective and sensitive analysis. This technique involves the simultaneous use of two mass spectrometers that are linked together; the first mass spectrometer operates as a filter for the second, where detection and quantification are made. Harvan et al. (24) achieved an absolute detection limit of less than 1×10^{-12} g for dioxin by using high-resolution MS–MS. Davidson et al. (25), using GC–MS–MS, reported a sensitivity of 20 parts per trillion for dioxin in environmental samples.

Analysis Versus Risk Assessment

Thus, in the span of about two decades, detection limits have been reduced about 6 orders of magnitude. These advances in analytical methodology have allowed scientists to study the movement of pesticides through the environment and the food chain, have been invaluable in the determination and quantification of metabolic and environmental degradation products, and have allowed federal and state regulatory agencies to monitor food and water. The advances in analytical methodology have also played a significant role in the registration of pesticides designed to have minimal environmental effects.

Significant advances in biochemical toxicology have also occurred since publication of *Silent Spring*, and some fundamental toxicological mechanisms are well understood. The connection between chemical carcinogens and chemical mutagens has become well-established, and determination of such agents has become much easier because of the development of numerous short-term assays. Carson wrote, "Although chemical manufacturers are required by law to test their materials for toxicity, they are not required to make the tests that would reliably demonstrate genetic effect and they do not do so." Happily, this situation has changed. Mutagenicity and carcinogenicity data for new pesticide registrations are now required.

Occasionally, however, pesticides that were registered before the advent of short-term mutagenesis assays and before changes in protocols for carcinogenic evaluation were made are discovered to be animal carcinogens. In such cases, the divergence between our ability to detect and quantify at extremely low levels and our inability to adequately assess human risk leads to serious problems.

Because animal carcinogenesis assays are insensitive, testing is carried out at the highest dose that the animals can tolerate. This dose is known as the *maximum tolerated dose* (MTD) and is commonly referred to as the high dose. The low dose is one-half the MTD.

Figure 2 (26) depicts the results of a hypothetical animal chemical carcinogenesis experiment. At the doses tested, a straight-line (linear) relationship occurs between dose and response. That is, the higher the dose, the greater the response.

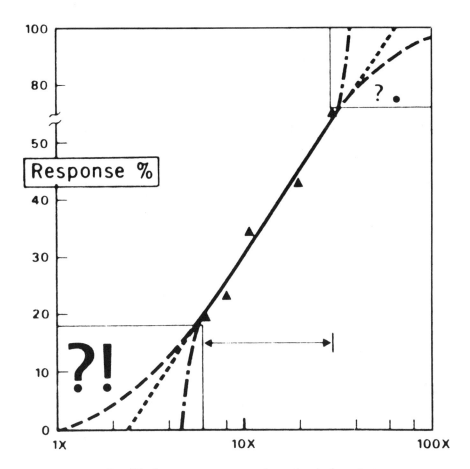

Figure 2. Possible dose–response curves for a chemical carcinogen. (Reproduced with permission from reference 26. Copyright 1978 American Association for the Advancement of Science.)

Unfortunately, experiments with more than 50 to 100 animals in each test group are very difficult to perform, and a carcinogenic response 20% greater than that of the control group is needed for reliable statistical analysis with such relatively small groups of animals. Thus, the shape of the curve at the doses to which humans are generally exposed (and the levels that the analytical chemist can easily measure) is unknown. If the dose-response curve were to remain linear or become supralinear, the line would pass through the x-axis and a threshold value could be determined. A *threshold value* is the level at which a response occurs and below which, no response occurs. However, if the curve were to become sublinear, there would be no threshold value and a finite carcinogenic response could be expected at all dose levels above zero.

The question of a threshold value for chemical carcinogens has been debated by scientists for many years and will probably never be determined scientifically because of the tremendous numbers of laboratory animals needed to answer the question. Carson, for one, did not believe that a threshold value for a carcinogen existed. Many other scientists, however, believe that detoxification and repair processes are overwhelmed at the high doses that the test animals are subjected to and that threshold values for carcinogens must exist. Regulatory agencies have taken the position that, in the absence of conclusive data to the contrary, threshold values do not exist for carcinogens, and these agencies rely on mathematical models to determine "safe" levels of exposure. Currently, five mathematical models can be used to extrapolate the results of laboratory experiments to low doses. On the basis of the assumption used in generating the models, "safe" doses can differ by as much as 3–4 orders of magnitude (27). Clearly, a pesticide registration may be revoked on the basis of inadequate data and selection of a particular mathematical model.

The recent regulatory decision (28) to prohibit the further use of ethylene dibromide (EDB) for soil, stored grain, and citrus fumigation illustrates the problems generated by our inability to accurately assess the carcinogenic risk of chemicals at exposure levels we can easily detect. EDB was registered for use in 1955 as a soil and commodity fumigant having nematocidal and insecticidal properties. Analysis methods for EDB were based on conversion to inorganic bromide (29) and measurement of bromide by the iodometric procedure of Kolthoff and Yutzy (30). This method was sensitive to 10 ppm of EDB expressed as total bromide. Advances in analytical methodology in the ensuing years lowered the detection limit to 1 ppb of EDB per se (31, 32). Confirmation of residue identity at this level can be made by a combination of electron-impact and negative-ion chemical ionization in the selected-ion monitoring mode (33).

In 1977, Weisburger (34) reported that EDB caused a high incidence of tumors in rats and mice. The doses used in these experiments were about 40 mg/kg/day in rats and 107 and 62 mg/kg/day in mice. [Two groups of rats (one male and one female) were tested at about 80 mg/kg/day, but this dose had to be cut back considerably because of very high toxicity.] In order to err on the side of safety, EPA used the 40-mg/kg/day rat dose for dietary risk assessment because it was lower than the doses administered to mice. From both analytical and human grain consumption data, the agency was able to calculate an average EDB daily intake for humans of 5.5 ng/kg/day (35), a dose approximately *7 orders of magnitude lower* than the dose administered to rats. Somehow, EPA was able to calculate an increased dietary risk of 1 per 1000 due to EDB use on the basis of the consumption data, *one rat data point*, and the Wiebull mathematical model (28). This calculation has little underlying science to back it up. Unfortunately, numbers lend credence to regulatory decisions, and we will be given numbers no matter how these numbers are obtained.

Glossary

Halogen   The elements fluorine, chlorine, bromine, and iodine.

Halogenated pesticides   Pesticides containing halogens.

Micrograms per liter (μg/L)   Parts per billion.

Milligrams per liter (mg/L)   Parts per million.

Partitioning   Separating the components of a mixture by taking advantage of their preferences for one solvent over another.

Acknowledgments

This chapter is New Jersey Agricultural Experiment Station Publication No. F-10201-1-86.

Literature Cited

1. Schechter, M. S.; Soloway, S. B.; Hayes, R. A.; Haller, H. L. *J. Ind. Eng. Chem. Anal. Ed.* **1945,** *17,* 704–709.
2. Blinn, R. C.; Gunther, F. A. *J. Agric. Food Chem.* **1955,** *3,* 1013–1016.
3. Ordas, E. P.; Smith, V. C.; Meyer, C. F. *J. Agric. Food Chem.* **1956,** *4,* 444–453.
4. O'Donnel, A. E.; Johnson, H. W.; Weiss, F. T. *J. Agric. Food Chem.* **1955,** *3,* 757–762.
5. Jones, L. R.; Riddick, J. A. *Anal. Chem.* **1952,** *24,* 569–571.
6. Frawley, J. P.; Laug, E. P.; Fitzhugh, O. G. *J. Assoc. Off. Agric. Chem.* **1952,** *35,* 741–745.
7. Mills, P. A. *J. Assoc. Off. Agric. Chem.* **1959,** *42,* 734–740.
8. Mitchell, L. C. *J. Assoc. Off. Agric. Chem.* **1952,** *35,* 928.
9. Coulson, D. M.; Cavanagh, L. A.; De Vries, J. E.; Walter, B. *J. Agric. Food Chem.* **1960,** *8,* 399–402.
10. Burke, J.; Johnson, L. *J. Assoc. Off. Agric. Chem.* **1962,** *45,* 348–354.
11. Mills, P. A.; Onley, J. H.; Gaither, R. A. *J. Assoc. Off. Agric. Chem.* **1963,** *46,* 186–191.
12. Lovelock, J. E.; Lipsky, S. R. *J. Am. Chem. Soc.* **1960,** *82,* 431–433.
13. Burke, J.; Giuffrida, L. *J. Assoc. Off. Agric. Chem.* **1964,** *47,* 326–342.
14. Coulson, D. M. *J. Gas Chromatogr.* **1965,** *3,* 134–137.
15. Hall, R. C. *J. Chromatogr. Sci.* **1974,** *12,* 152–160.
16. Giuffrida, L. *J. Assoc. Off. Agric. Chem.* **1964,** *47,* 293–300.
17. Brody, S. S.; Chaney, J. E. *J. Gas Chromatogr.* **1966,** *4,* 42–46.
18. Kolb, B.; Bischoff, J. *J. Chromatogr. Sci.* **1974,** *12,* 625–629.
19. Biros, F. J. *Res. Rev.* **1971,** *40,* 46.
20. Cairns, T.; Siegmund, E. G.; Jacobson, R. A.; Barry, T.; Petzinger, G.; Morris, W.; Heikes, D. *Biomed. Mass Spectrom.* **1983,** *10,* 301–315.
21. Gordon, A. E.; Frigerio, A. *J. Chromatogr.* **1972,** *73,* 401–417.
22. Rosen, R. T.; Rosen, J. D.; DiProssimo, V. P. *J. Agric. Food Chem.* **1984,** *32,* 276–287.
23. Stout, S. S.; Steller, W. A. *Biomed. Mass Spectrom.* **1984,** *11,* 207–210.
24. Harvan, D. J.; Hass, R.; Schroeder, J. L.; Corbett, B. J. *Anal. Chem.* **1981,** *53,* 1755–1759.
25. Davidson, W. R.; Sakuma, T.; Gurprasad, N. *Anal. Chem.* **1983,** *55,* 762–764.
26. Maugh, T. H., II, *Science* **1978,** *202,* 37.
27. Krewski, D.; Brown, C.; Murdoch, D. *Fundam. Appl. Toxicol.* **1984,** *4,* S383–S394.

28. *Fed. Regist.* **1983,** *48(197),* 46235–46236.
29. Mapes, D. A.; Shrader, S. A. *J. Assoc. Off. Agric. Chem.* **1957,** 40, 189–191.
30. Kolthoff, I. M.; Yutzy, H. *J. Ind. Eng. Chem. Anal. Ed.* **1980,** 9, 75–76.
31. Clower, M. *J. Assoc. Off. Anal. Chem.* **1980,** 63, 539–545.
32. Rains, D. M.; Holder, J. W. *J. Assoc. Off. Anal. Chem.* **1981,** 64, 1252–1254.
33. Barry, T.; Petzinger, G. *J. Food Safety* **1985,** 7, 171–176.
34. Weisburger, E. K. *Environ. Health Perspect.* **1977,** 21, 7–16.
35. Burin, G. Presented at the 188th National Meeting of the American Chemical Society, Philadelphia, PA, August 1984; paper ACSC 7.

9 ∾ Pesticides: Global Use and Concerns

Virgil H. Freed

The publication of *Silent Spring* in 1962 fanned the smoldering concern in the United States over the use of pesticides into a flame of action. The book prompted the formation of new environmental organizations and stimulated old organizations to question the use of certain chemicals. Recommendations were developed by organizational, governmental, and National Academy of Sciences committees for stringent regulation or phase-out of certain pesticides (1–3).

Since the publication of *Silent Spring*, a spate of books, both pro and con, has been written on the same subject (4–9). In addition, the debate over the dangers and benefits of pesticides has raged with fervor in public, in legislative halls, and in courtrooms. The emotions evoked in this debate extend from a lack of quantitative information on both the risks and the benefits of pesticide use. More recently, the debate has evolved into a rational dialogue. But the complexities and compounding factors of these problems and the lack of quantitative data will probably limit our ability to provide definitive answers for some time. The concern over pesticide use reaches not only developed countries but developing countries as well.

We have been practicing agriculture for more than 10,000 years to produce food needed for existence. Pests have always been keen competitors. Hence, rapid, effective control of these

0980–4/87/0145$06.00/0 © 1987 American Chemical Society

pests is an attractive prospect to the practicing agriculturist. Unfortunately, as Rachel Carson so forcefully pointed out in her book, pesticides were often overused and sometimes misused in the early years of the era of modern organic pesticides.

Chemicals Used in Agriculture

Beginning about 3000 years ago, philosophers and others in Chinese, Greek, and Roman societies developed an interest in chemicals available for use as elixirs, potions, and medicines. Subsequently, they discovered that some of the elements, such as sulfur, were effective in controlling plant diseases (9). Little was made of this observation for many hundreds of years. As late as the mid-1700s, only a few people on this continent were using infusions of tobacco for control of insects.

As the sciences and technologies developed, materials such as copper salt and saltwater were found to be effective inhibitors of seed-borne diseases of cereals. Later, the efficacy of certain chemicals, such as lead arsenate, in killing insects was observed, and the knowledge was put to use.

Approximately 1500 different chemicals go into the slightly more than 1.5 billion pounds of pesticides manufactured annually in the United States. About 65 chemicals make up the major portion of that amount (10). Several thousand compounds have demonstrated biological activity for control of bacteria, fungi, insects, plants, and rodents; and several thousand more are used in the processing and preservation of food.

Pesticide Use

Reasons for Use

The use of pesticides for health protection and agricultural production is not without risk. If the benefits were not assured, many pesticides would not be used. To be sure, routine application of pesticides shows little regard for assessment of benefits or risks to humans and the environment.

Careful use of pesticides can result in health protection from vector-borne diseases (11). For example, malaria in the tropical countries was substantially abated for many years through the use of pesticides. However, the development of resistant strains of mosquitoes in some instances has severely constrained the effectiveness of the chemical control program (*12*). Integrating other techniques with the use of chemicals is proving helpful.

Control of agricultural pests through the correct use of insecticides, herbicides, and fungicides has contributed significantly to increased food production (*13*). Adequate production of food is not a problem in many of the developed countries. However, in the developing nations food production is a critical problem. For example, in parts of Africa the per capita food production has actually declined (*13*). Many countries, recognizing the critical food problem, are giving greater emphasis to increasing their agricultural yield (*12*). Along with improved crop varieties, better management, and increased use of fertilizers, pesticides play an important role in increasing food production.

Kinds of Pesticides

The classes of chemicals use as pesticides have changed dramatically since the publication of *Silent Spring* in 1962. At that time the chlorinated organic compounds (DDT was the prime example) were the most widely used pesticides. Although the organophosphates (e.g., malathion and parathion) had come into use by then, the greatest criticism was leveled against the use of chlorinated organic chemicals. The concern over this class of chemicals was based on their refractory nature and lipophilicity that made them persist in the environment; this persistence resulted in their transport by wind and water and their bioaccumulation (*3, 14*). When chlorinated organic compounds were used in a careless manner, these properties, coupled with the biological activity of the chemicals, caused kills of aquatic organisms and reproductive difficulties in birds (*14*).

The chlorinated organic compounds were phased out of use for three reasons: (1) restrictions imposed by the U.S. Environmental Protection Agency (EPA), (2) the development of more effective and less persistent compounds, and (3) the growing number of species resistant to the chlorinated hydrocarbons (*14*). New classes of chemicals were introduced as pesticides.

Those classes used as insecticides included organophosphates, N-alkylcarbamates, synthetic pyrethroids, formamidine derivatives, and several other types of chemicals. Materials introduced as herbicides and fungicides were equally diverse. Herbicides were represented by such chemical classes as ureas, triazines, uracils, benzoic acid, pyridine derivatives, and organophosphates.

Although the organochlorine pesticides are persistent and therefore chronically hazardous to wildlife, when properly applied most of them afford only a small risk to the user. In contrast, many of the organophosphates and carbamates used as insecticides have a high toxicity to humans and other mammals. These compounds now are a common cause of poisoning among pesticide workers (10).

Level of Use

Despite the continuing and spreading controversy over the use of pesticides, the use of such chemicals has increased since the publication of *Silent Spring*. The average annual increase in pesticide use has been about 4 to 5% on a global basis (15, 16). The current level of use is estimated to be more than 4 billion pounds of active ingredient annually (17).

The developed nations in Western Europe and North America use about 57% of the annual sales of pesticides, Latin America uses about 15%, Eastern Europe uses about 11%, Japan uses about 11%, and the rest of the world uses the remaining 6% (18). The amount of pesticide active ingredient actually imported or manufactured and used in developing countries is difficult to ascertain. The estimated import of pesticides by Malaysia is given as 16 million liters for 1976 (19), but no concentration is provided by which to estimate the pounds or kilograms of active ingredient. Investigators in Costa Rica (18, 20) have also tried to obtain estimates of the amount of pesticide imported.

Transport and Modeling of Pesticides

Perhaps the most intensive attempt to model transport and behavior of a pesticide in the total environment was done with

DDT. Harrison (21) provided one of the early detailed models of the behavior of DDT in an attempt to estimate transport through various trophic levels (i.e., steps in the food chain) and possible persistence of the chemical. Woodwell (22), in assessing a global model for transport and persistence of DDT, found the concentration much lower in certain compartments (i.e., the atmosphere, soil, or water) than would be predicted on the basis of the model. Kramer (23) attempted to give a quantitative description or model of the circulation of DDT. He pointed out that rates of degradation, transfer, and adsorption by plankton are critical factors in such a model. Monitoring data suggest that the rate of disappearance of DDT may be more rapid in the general environment than previously estimated. Others have given attention to modeling behavior of pesticides other than the organochlorines in the atmosphere, soil, and water. As yet, insufficient data have been accumulated to develop a fully consistent global model, although such models as EPA's EXAMS (exposure analysis modeling system) have proven very useful to estimate the probable fate and behavior of chemicals.

The principal mechanisms of transport of chemicals are wind (14, 24–26), water, and biota. After introduction into the environment, chemicals may become available for one or more of these transport mechanisms by a process such as vaporization. Thus, the chemical may become available for wind transport during spraying as the carrier evaporates and leaves small droplets or particles of the pesticide suspended in air. If the chemical evaporates either from the droplet or from a surface, it can be transported just by air movement. Or erosion of soil contaminated by the chemical may result in wind transport as demonstrated by Cohen and Pinkerton (27).

Similarly, water transport may be a result of soil erosion, leaching, or direct entry into the waterway (22, 26, 28). Wauchope (26) reviewed the percentage losses of a number of pesticides from runoff. He found that the data varied depending on water solubility, pesticide formulation, soil characteristics, slope, and intensity of rainfall. Biotic transport may be the result of migrating populations of organisms (e.g., birds and fish) contaminated with the chemical through their food (22, 23). More importantly, biotic transport can occur because of human activity.

Wind Transport

Mass air movement is an important means of transport of low levels of chemicals to distant points (14, 21, 23, 27, 28). This subject has engaged the interest of meteorologists and air pollution scientists for many years. From their studies, we can learn much that is applicable to the problem of transport of pesticides.

Factors relating to air transport include rates of vaporization, surface area exposed, quantity of chemicals available, rate of photodecomposition, fallout, and rainfall (14, 28, 29). For many of the organic compounds, the residence time in the atmosphere varies from a few hours to a few weeks depending on their resistance to decomposition. The limited evidence available suggests that many pesticides have a rather short-term residence in the atmosphere.

As would be anticipated, the concentration of the chemical in the atmosphere and subsequently on surfaces decreases as the distance from the source increases. Dilution and decomposition as well as fallout contribute to the decline in concentration.

Water Transport

The problem of water transport, particularly that rising from runoff in which chemical or contaminated particulate matter is carried into bodies of water, is receiving more attention today (26, 30). The solubility and adsorption characteristics of the chemical are very important in analyzing this process. Thus, the higher the water solubility, the greater the proportion of chemical found in the runoff water (26). The chemicals of greater adsorption and lower water solubility, when found in runoff water, are usually associated with particulate matter. In this case, the mathematical distribution coefficient such as the one used extensively by Hamaker (31) and others determines the relative distribution of the chemical between the aqueous and the particulate phases.

Biotic Transport

Biotic transport by a migrating population probably does not involve large quantities of chemical (14). If, however, the migrating population is in the food chain of another organism, the chemical may be concentrated in a few organisms at a distant point. To put the matter in perspective, if a migrating population of 1 million kg (total weight of the population) carried 10 parts per million of residue, the population would transport only 10 kg of the chemical. However, the same level of residue might bioaccumulate in a population having a mass of 100,000 kg; this situation would give a 10-fold increase in apparent concentration. This assumption is very idealized because we know that the transfer mechanism (bioaccumulation) is not 100% efficient. It is more likely about 25% efficient and gives only a 2.5-fold bioaccumulation factor.

Humans transport chemicals in other ways. A chemical used as a pesticide at Point X on a food crop can be transported to Point Y on that crop, but in most instances the residue level is significantly reduced in the processing and preparation of the food by washing and cooking. We manufacture technical chemicals at one point and transport them to another point for use. In this transport, appropriate precautions can be exercised to avoid accidents and adverse effects.

Problems Encountered with Pesticides

As a result of observation of pesticide use in a number of countries, four problems were described (*10*):

1. human and animal poisoning

2. residues in foods in the environment

3. the development of resistance, particularly on the part of insects

4. the problem of safe disposal of pesticide wastes and containers

Perhaps a fifth problem should be added to this list: the development of rapid loss of effectiveness of some of the soil-applied chemicals probably because of microbiological activity.

These problems are of equal concern for both health and agriculture. Improper use of a chemical in either field is likely to impinge on the other. This situation is illustrated by the fact that use of certain insecticides in some countries has resulted in the development of resistant strains of insect vectors (12).

Human and Animal Poisoning

When agriculture and public health services switched from the organochlorine to the organophosphate and carbamate insecticides, human and animal pesticide poisoning increased worldwide (15, 32). Human poisonings from organochlorine compounds did occur, but the organophosphates, being particularly more toxic to mammals, greatly exacerbated the problem. The paucity of statistics on poisoning shows that the magnitude of this problem is still not fully known because, in many areas of the world, acute pesticide-related incidents are often unrecognized and underreported.

In a review of global pesticide safety, Copplestone (15) discussed the World Health Organization (WHO) estimates of the magnitude of the problem of pesticide poisoning. The WHO expert committee on the safe use of pesticides concluded (33) in 1972 that, on the basis of the mathematical model based on the statistics of accidental poisoning for 19 countries, 500,000 pesticide poisoning cases occur annually. The mortality rate was estimated to be about 1% in those countries where medical treatment and antidotes were readily available. In a subsequent survey of a series of different countries, a poisoning rate of 2.9 to 4.8 per 100,000 persons was found (10). The mortality rate, based on the number of poisoning cases, was about 5.5%; in other words, 1 out of about every 18 cases of poisoning resulted in death. In many instances, the poisoning was the result of contamination of food or clothing as well as careless use of the chemicals.

Residues and Environmental Pollution

The contamination of water supplies, both surface and ground water, is a significant problem that is developing from environmental pollution by pesticides and other chemicals. Continued application of pesticides, particularly the more water-soluble types, can result in their transport to streams and other water sources or in leaching through the soil profile to ground water that may be used for drinking.

Persistence of chemicals contributes to residues in both food and the environment. The problem is especially acute with the organochlorine pesticides, many of which continue to be used in tropical agriculture. However, even the organophosphates, when applied too near to harvest or at too high a rate, constitute a problem.

The problem of residues in food and the environment is illustrated by a case I observed in Central America. In the production of cotton, a preharvest application of organochlorine was a common practice. After harvest, beef cattle were allowed to graze on the cotton stalks before being taken to fattening pens. In the fattening pens they were fed cottonseed meal and corn that had been treated with an organochlorine insecticide for earworm control. Needless to say, the residue levels in the beef were significantly high.

Development of Resistance

The speed of the biological clock in tropical areas means that more generations of a given organism will be found there than in temperate regions. Therefore, is is not surprising to find that, because multiple applications of pesticides are needed to control pests, resistance has developed, again primarily in the insects but also in some tropical weeds. The resistance of insect vectors is often fostered by the use of pesticides in agriculture (10, 12). For example, the extensive use of organochlorine and organophosphate pesticides on cotton helps vectors of human disease that are indigenous to the area develop resistance. Resistance, however, is not always confined to the chemical used; organisms may show cross-resistance, as in the case of mosquitoes that have developed resistance to organophosphates and are also resistant to carbamates.

Disposal of Chemicals and Containers

The disposal of chemical wastes and containers is a problem of some magnitude in many tropical countries. High temperatures and humidity accelerate the decomposition of pesticides in open containers. Thus, distressed stock must often be disposed of. In the case of government programs, the amount of chemical ordered frequently exceeds the requirements. The remainder is poorly stored more often than not, and deterioration of both chemical and container results. All too frequently, disposal consists of dumping the material into a nearby stream or garbage dump. It takes little imagination to envision the problem that ensues (4, 10).

Concerns over Pesticide Use in the Third World

Some of the problems resulting from the use of pesticides in Third World countries have aroused considerable alarm. The developed countries, who are the primary manufacturers of pesticides, have been accused of exporting poisons (4, 34). Concern is expressed not only for the impact of the pesticides on the people and environment of the Third World countries but also that the residues of these materials in food crops may be exported to us.

The concerns expressed by some in the developed countries have influenced the Third World countries. In some instances, this concern has stimulated more awareness of pesticide problems, both current and potential, and has led to efforts to ameliorate these problems. In other instances, fear and over-reaction have resulted.

In most of the developed Western countries, life-threatening, vector-borne diseases are limited, and an adequate food supply is available year-round (35). Advanced health care and agricultural technology, including the use of pesticides, go far in ensuring this happy situation. Under these circumstances, evaluation of the use of pesticides can be more concerned with the possible effects from long-term, low-level residues rather than with human survival.

Environmental conditions in most of the temperate-zone developed countries also are substantially different than in the developing countries. For example, the climate of the countries in the humid tropics fosters the multiplication of ectoparasites and vectors of human and animal diseases and advances the biological clock of crop pests. Multiple generations, therefore, constitute a year-round threat. In many of these countries, 50 to 90% of the economy is based on agriculture. Much of the agriculture is directed toward production of export crops in a plantation system (10, 13).

In both rural and urban areas, pesticides are relied upon as an immediate and effective way to control pests and diseases. This reliance is reflected in the large increase in the amount of pesticides being imported by and sometimes manufactured in tropical countries. For example, since about 1965, the use of pesticides in Africa has increased almost fivefold; similar increases are reported for other areas of the world (36).

Even the safest pesticides are often used in a manner that threatens the safety of the user and the environment. For example, in 1976 in the malaria control program in Pakistan, 2900 people were poisoned by malathion. Five of these people died (37). In some areas, persistent pesticides, notably the organohalogens, continue to be used in agriculture.

As noted earlier, environmental conditions in the tropical countries are substantially different than those in the temperate region. Higher temperatures and, in many instances, higher moisture levels combined with more intense irradiation cause more rapid loss of chemicals through degradation and volatilization (38, 39).

Furthermore, Third World countries are often short on technically trained people. The number of people who are well trained in pesticide use is small. Moreover, in many instances, the populace has not been exposed to the advanced level of technology taken for granted in the West. As a consequence, they are less familiar with the safe use of pesticides as technical tools in agriculture and health care. However, it must be remembered that the area treated or the extent of pesticide use in developing countries is usually small in proportion to the overall land area. Heavy use through frequent application on a limited area does occur, but it would not involve as high a proportion of cultivated acreage as in developed countries.

Glossary

Active ingredient   The pesticide or ingredient that brings about the desired biological action.

Chlorinated organic compound   A compound made up of carbon, hydrogen, and chlorine. It may occasionally contain other elements.

Compartments   Soil, water, air, or one or more of the biological species in the environment.

Distribution coefficient   The ratio of concentration between two separate compartments or phases, for example, between soil and water.

Fallout   Material redeposited, usually from the atmosphere; often used to describe deposit of radioactive material and may also apply to pesticides.

Global model   A quantitative or qualitative conceptualization to show relationships.

Leaching   Movement usually through soil by water.

Lipophilicity   Tending to be soluble in fats.

Organophosphates   A class of pesticides containing phosphorus.

Photodecomposition   The breakdown of a chemical by light, particularly the ultraviolet end of the spectrum.

Refractory nature   Resisting change.

Trophic level   A level in the environment or biota, for example, in a food chain, the various levels up to and including the ultimate consumer.

Vector   An insect or an agent carrying and transmitting a disease or another biological agent.

Literature Cited

1. *Persistent Pesticides;* National Academy of Sciences: Washington, DC, 1980.
2. *Regulating Pesticides;* National Academy of Sciences: Washington, DC, 1980.
3. Report to Secretary (Mrak Report). U.S. Department of Health, Education, and Welfare. U.S. Government Printing Office: Washington, DC, 1969.
4. Bull, D. *A Growing Problem: Pesticides and the Third World Poor;* OXFAM: Oxford, 1982.
5. Graham, F. *Since Silent Spring;* Houghton Mifflin: Boston, 1970.
6. Green, M. B. *Pesticides: Boon or Bane?;* Westview: Boulder, CO, 1976.

7. Hay, A. *The Chemical Scythe;* Plenum: New York, 1982.
8. Lappe, F. M.; Collins, J. *Food First;* Houghton Mifflin: Boston, 1977.
9. Whitten, J. L.; *That We May Live;* D. Van Nostrand: Princeton, N.J., 1966.
10. *An Agromedical Approach to Pesticide Management: Some Health and Environmental Considerations;* Davies, J. E.; Freed, V. H.; Whittemore, F. W., Eds.; University of Miami Press: Miami, 1982.
11. Southwood, T. R. E. *Am. Sci.* **1977,** *65,* 30.
12. Chapin, G.; Wasserstrom, R. *Nature* **1981,** *293,* 181.
13. Mellor, J. W.; Adams, R. H., Jr. *Chem. Eng. News* **1984,** *Apr. 23,* 32.
14. Brown, A. W. A. *Ecology of Pesticides;* Wiley: New York, 1978.
15. Copplestone, J. F. In *Pesticide Management and Insecticide Resistance;* Watson, D. L.; Brown, A. W. A., Eds.; Academic: New York, 1977.
16. Darmansyah, I. In *Pesticide Management and Insecticide Resistance;* Watson, D. L.; Brown, A. W. A., Eds.; Academic: New York, 1977.
17. *Formulation of Pesticides in Developing Countries;* Maier, A.; Zweig, G., Eds.; United Nations Industrial Development Organization (UNIDO), United Nations: New York, 1983.
18. Maltby, C. *Report on the Use of Pesticides in Latin America;* United Nations Industrial Development Organization (UNIDO), United Nations: New York, 1981.
19. Sahabat Alam *Pesticide Problems in a Developing Country—A Case Study of Malaysia;* Sam, Penang: Malaysia, 1981.
20. Vega, S. et al. *Importacion Y Exportacion de Plaguicidas En Costa Rica;* Universidad Nacional Heredia Costa Rica, 1983.
21. Harrison, H. L. *Science* **1970,** *170,* 503.
22. Woodwell, G. *Science* **1971,** *174,* 110.
23. Kramer, J. *Atmos. Sci.* **1973,** *7,* 241.
24. Haque, R.; Kearney, T. C.; Freed, V. H. In *Pesticides in Aquatic Environment;* Kohn, N. A. Q., Ed.; Plenum: New York, 1976; p 39.
25. MacKay, D.; Wolkolff, A. W. *Environ. Sci. Technol.* **1973,** *7,* 611.
26. Wauchope, R. D. *J. Environ. Qual.* **1978,** *7,* 459.
27. Cohen, J. M.; Pinkerton, C. In *Organic Pesticides in the Environment;* Advances in Chemistry 60; American Chemical Society: Washington, DC, 1966, p 63.
28. Taylor, A. W. *J. Air Pollut. Control Assoc.* **1978,** *28,* 922.
29. Glotfelty, D. E. *J. Air Pollut. Control Assoc.* **1978,** *28,* 917.
30. Pavlou, S. P.; Dexter, R. N. *Environ. Sci. Technol.* **1979,** *13,* 65.
31. Hamaker, J. W. In *Environmental Dynamics of Pesticides;* Hague, R.; Freed, V. H., Eds.; Plenum: New York, 1975; p 115.
32. *National Study of Hospital-Admitted Pesticide Poisoning.* U. S. Environmental Protection Agency. U.S. Government Printing Office: Washington, DC, 1976.
33. Technical Report No. 513, 1972, World Health Organization, p 42.
34. Weir, D.; Shapiro, M. *Circle of Poisons: Pesticides in People in a Hungry World;* Institute for Food Developments: San Francisco, 1981.

35. Claus, G.; Bolander, K. *Ecological Sanity;* David McKay Co.: New York, 1977.
36. *Chemistry of Pesticides,* Buchel, K. H., Ed.; Wiley: New York, 1983.
37. Baker, E. L. In *Pesticide Conspiracy;* Robert VandenBosch, Ed.; Prism Press, 1977.
38. Sleicher, C. A.; Woolgrath, G.; Presented at the American Chemical Society-Chemical Society of Japan Joint Meeting, Honolulu, HI, 1979.
39. Talekar, N. S., Asian Vegetable Research Development Center, Taiwan, 1981, personal communication.

10 ∽ Agriculture, Pesticides, and the American Chemical Industry

Gustave K. Kohn

To assess the effects of *Silent Spring* on industry, and particularly the agrochemical industry, we will evaluate not what representatives of industry have said or written, but what industry did, is doing, and will do. This evaluation must include an examination of the substances (quantities manufactured, physical and chemical properties, environmental impacts, etc.) produced before 1962 and since then.

Another part of this assessment of the effects of *Silent Spring* transcends science. It deals with attitudes and values, and it has many social and political implications.

The Demographic Nature of American Agriculture

Until 1800, approximately 90% of the American population was primarily associated with the growing of food and fiber. Roughly at the period of the Civil War, the urban population equaled the farming population. By the 1960s only about 5% of the population supplied all the food needed for consumption by our nation, and we were feeding much of the world just recovering from the ravages of World War II. Agricultural production was the brightest ornament of the American

0980–4/87/0159$06.00/0 © 1987 American Chemical Society

economic landscape. We still are most envied by the Iron Curtain countries and most of the world for our agricultural achievements. Today, only 3-4% of our population supplies food and fiber cheaply and abundantly. Approximately 20% of the U.S. population is engaged not in the actual farming but in the industries involved in the collection, transport, warehousing, distributing, and processing of materials derived from farm production. This demographic fact has many consequences:

- We have departed from the Jeffersonian ideal. Most Americans live in the cities. We are no longer a nation of free and independent producers of wealth.

- The great majority of Americans have been liberated from having to labor in the fields for sustenance. Edwin Markham's poem and Francis Davis Millet's painting illustrate the changes wrought by technology.

- The political power of the farm versus the urban population is very small. Only if large sections of the urban population understand and act upon the knowledge that food is primary to the economy as a whole can we avert serious political mistakes.

- The migration of poorly educated sections of the population to the cities (e.g., deprived blacks from the agrarian South to the cities of the North), has resulted in serious social and economic dislocations.

The agrochemical industry today and the demographic distribution are the consequences of the application of technology to agriculture. The American chemical industry is no more responsible for these consequences than the automobile, publishing, or textile industries.

A Conflict of Values

The desirability of high agricultural productivity cannot be disputed. We enjoy an economy where the cost of food is low relative to the citizens' total budget, and the supply is plentiful. This cost advantage (which is presently eroding) was achieved by the ever-increasing application of technology to farming. The American tradition had resulted in legislation throughout the 19th and early 20th centuries that culminated in the land-grant colleges and the extension system to make contemporary science and technology the special property and the privilege of the American farmer. This science and technology included tremendous advances in classical genetics and plant breeding; the invention and application of all sorts of mechanical devices (such as tractors, threshers, and planters); and the utilization of fertilizers and chemicals to provide nutrition, destroy parasites, and treat plant diseases, among other functions. American industry provided the means for these advances.

Pesticides do contribute to agricultural productivity. A current Soviet publication (*1*), although based on American studies, is cited to refute a prevalent delusion that the beneficial consequences of pesticide use constitute a capitalist industry ploy. Ten different crops were studied, with and without chemical treatment. In all cases, crop losses were reduced significantly with chemical treatment. The reductions ranged from 17% for tomatoes to 78% for apples and pears. The increase in yield as a result of chemical treatment ranged from 1.3 times greater for strawberries to 12 times greater for apples and pears. After the 1960s, particularly with the trend toward monoculture, the need for pesticides expanded.

From the period ending with World War II to the approximate time of the publication of *Silent Spring*, American agriculture experienced large increases in productivity. In those 15 years, corn production increased from 36.5 to 68.1 bushels per acre; wheat, from 16.0 to 25.9 bushels per acre; potatoes, from 137.6 to 198.0 hundred weight per acre; cotton lint, from 279.6 to 499.8 pounds per acre; and rice, from 2211.6 to 4073.8 pounds per acre. Dairy cow production (milk) increased from 5194.0 to 8022.0 pounds per cow per year; and eggs, from 169.0 to 216.0 per hen per year. Higher productivity is an accepted and valuable objective for industry as a whole and for agriculture.

The economist uses the term *externalities* to include the hidden costs that are usually left uncalculated and that frequently include ultimate costs to society and to the producers. Rachel Carson discusses these externalities in *Silent Spring*: the costs to the land and to the environment. The productivity value is expressed by a social critic of another century, Jonathan Swift (2):

> And he gavest for his opinion, that whoever could make two ears of corn or two blades of grass to grow upon a spot of ground where only one grew before, would deserve better of mankind, and do more essential service to his country than the whole race of politicians.

Three to four ears of corn now derive from that same spot of ground where only one grew in the 1930s.

Carson emphasized the neglect of the effects of the chemicals being used to achieve greater productivity on nontarget animal species, on the land, on human beings, and on the environment as a whole. These values also are significant. They are in conflict with the value of high productivity as an isolated value. The resolution of the conflict in values involved and involves American politics from 1962 to the present.

Pesticides Produced by American Industry Before *Silent Spring*

From the 1930s to the end of World War II, sulfur, oil sprays, and inorganic chemicals dominated the agrochemical arsenal. Production of inorganic chemical pesticides increased steadily throughout the 1930s. For example, production of calcium arsenate increased from 26 million pounds to 43 million pounds by 1935, then tapered off to 37 million pounds in 1937. Lead arsenate production increased from 37 million pounds to 63 million pounds, and pyrethrum from 4.7 million pounds to 7.1 million pounds by 1937. During the 1930s, tillage by manual labor was the predominant method of weed control. No wonder the farmer and agricultural experiment stations accepted the fruits of the synthetic organic chemical revolution that began during World War II. The new organic chemicals were enormously more target effective and apparently were, particularly the chlorinated hydrocarbons in general, much less acutely toxic to humans. In addition to their great effectiveness, many of these organic chemicals were extremely simple to manufacture. By 1950, organic chemical pesticides accounted for 60% of the total pesticide market, and by 1955, they were 90% of the total market.

While inorganic pesticides such as lead arsenate and calcium arsenate showed decreases or only modest increases in production during the 1950s, DDT (an organic compound) production increased from 37 million pounds in 1953 to 124 million pounds in 1959.

Of the three categories of pesticides (insecticides, fungicides, and herbicides), insecticides dominated the pesticide industry from the 1940s through the 1960s. Herbicide production had lagged behind until the mid-1950s, when it rose spectacularly. By 1960, herbicide production had surpassed fungicide production and equaled about one-fifth of the pesticide market.

Pesticides Produced at the Time of Publication of *Silent Spring*

In the 1960s the chlorinated hydrocarbons dominated the pesticide market. They were broadly effective, they persisted in the environment, and some had relatively low acute human toxicity. These were viewed almost as panaceas for farm application and public health programs, as reported in the *Journal of Economic Entomology* in the late 1940s through the 1950s. A majority of papers dealt with the beneficial effects of the application of insecticidal chemicals.

DDT was the most frequently used chlorinated hydrocarbon. It possesses good oil solubility and water insolubility. It concentrates after absorption or ingestion into fatty tissue. It is stable (though vulnerable enzymatically and photochemically). It persists in the environment, particularly in the cooler temperature-zone soils. It concentrates in nontarget animal species, especially those at the head of particular ecological chains. DDT was manufactured in the United States for more than a decade at quantities greater than 100 million pounds per year. (*See* the box "DDT and Its History".)

The mammalian toxic properties of the cyclodienes are more severe than those of DDT. The cyclodienes vary in persistence. They are more hazardous because photochemical reactions occur by two pathways with different photoproducts that lead to compounds even more toxic than the original pesticides. The mammalian toxicological effects of DDT perhaps have been exaggerated. Not so for the cyclodienes. The mechanism of action of these cyclodienes was never satisfactorily elucidated.

DDT and Its History

History

Originally synthesized in Zeidler, Germany	1874
Remained laboratory curiosity until Paul Mueller discovered insecticidal properties	1939
Nobel prize (medicine) awarded to Mueller	1948
Mass introduction into agriculture and public health programs	1949 onward
Development of trace analytical techniques; recognition of resistance	1950–1960
Recognition of residue; ecological and physiological properties; and their meanings	1960–1970
State and federal legislation severely restricted use	1969–1970
Use eliminated in United States and advanced countries; use limited in Third World countries	1970–1980

Pertinent Physical and Chemical Properties

- Very soluble in organic solvents, fats, and oil.

- Solubility in water at 22 °C is 1.2 parts per billion.

- Vapor pressure very low (1.7×10^{-7} mm 20 °), but forms distillate from moist surfaces; forms transportable aerosols that are relatively stable.

- "Steam distills" from water, moist soils, and animal hair.

- Originally thought to be very stable in soil and in biological media, as well as photochemically stable. Metabolic pathway studies, photochemical degradations, and soil losses indicate exaggeration of early estimates on stability. Nevertheless, DDT is a relatively stable and persistent pesticide.

- Environmental burden estimated as 1×10^9 lb (3).

Insights into their toxicological properties have only recently been published.

The usage of these halogenated chemicals and their properties stimulated the mind of Carson, and therefore, DDT and the cyclodienes are the unconscious "coauthors" of *Silent Spring* and the social and political events that followed.

Carson greatly merited high praise for alerting the public to the ever-increasing environmental burden of persistent chemicals, the injurious effects of many of these chemicals on nontarget species, and the effect of insect resistance that necessitated ever-increasing dosage rates.

The Newer Agrochemicals of the 1980s

Industry did respond to Carson's criticisms with improved chemical pesticides. Starting in the early 1960s, industrial agrochemical research was directed toward the production of less persistent, nonbioaccumulative chemicals and toward alternative approaches to farm and public health strategies. Some companies aimed at insect control on the basis of hormonal interference with metamorphosis. Other companies emphasized weed control (sometimes exclusively) and the creation of biochemically site-specific methods of plant disease control. The use of agrochemicals did not decline. It increased, but it is now leveling off for all uses including herbicides. This leveling off and probable future decline will be the result of the much greater efficacy of the new pesticides that supplant the pesticides of the 1960s. The pyrethroids are the most-used insecticides in the 1980s. DDT was used at 1 to 5 pounds per acre, but pyrethroid dosage varies from 0.1 to 0.01 pounds per acre. Certain of the newer herbicides are and will be used at the rate of a few ounces per acre. The new Dupont and American Cyanamid branched amino acid inhibitor herbicides are applied in grams per acre, not pounds per acre.

In addition to using smaller doses of pesticides, less frequent application is now common. Spray programs are no longer prophylactic and automatic with the season, but are used as part of integrated pest management (IPM) programs only when dangerous infestation levels are reliably perceived.

An example of progress in the 1980s is the employment of insect growth regulators (IGRs) in the prophylactic population control of one of nature's evolutionarily most successful insects: the cockroach, ancient even when the dinosaurs were just beginning to emerge. Formerly many highly toxic substances, including persistent chlorinated hydrocarbons, organic phosphates and carbamates, and other hazardous substances, were employed in the home and elsewhere to control these pests. These IGRs are innocuous to mammals and provide population elimination (insect birth control) rather than insecticidal action.

For agriculture, current industrial R&D emphasizes biological approaches to improve production. One such approach is the use of products from fermentation methods, for example, microorganisms such as *Bacillus thuringiensis* that infect insects. Most agricultural chemical companies are currently working on genetic engineering methods to provide crops of the future that will be resistant to insect depredation and disease. Some companies work exclusively in this domain, and others make advanced biological methodologies their major R&D effort.

America's role in the manufacture and export of pesticides is declining. Export dominance is currently with Japan and Western Europe. Manufactured agrochemicals will not greatly increase and will probably decline, at least until such time that Third World economics vastly improves.

Carson influenced all of the changes just described. American industry independently and in response to her challenge is now engaged in scientific R&D that no one in the 1960s would have reasonably envisaged.

Technology and Farm Income

In 1941 only 4% of the corn crop was estimated to have been lost to insect depredation, but in 1983 12% of the corn crop was consumed by insects. These estimates were from the U.S. Department of Agriculture. Figures such as these are sometimes cited as evidence of the increasing costs of raising a corn crop and, by implication, a failure of modern insecticidal strategies. However, this conclusion is simplistic and unwarranted.

Infestation by insects (also by fungi, bacteria, weeds, etc.) varies from year to year because of a variation of environmental factors and factors unique to the organism. Therefore 5-year averages are more representative figures. For 1941–1945, the average yield of corn was 31.9 bushels per acre, and for 1978–1982, it was 105.4 bushels per acre, a 3.3-fold increase.

This increase is a result of the application of all kinds of technologies to corn production. Prominent among these are genetics, fertilization, harvesting practices, and pesticides. The new corn varieties were selected for yield and not for insect resistance. Increased quantities of fertilizer are employed to provide optimum plant nutrition. Crops that are well fertilized, particularly with nitrogen, have increased aphid infestations. Speaking anthropomorphically, the insect finds the well-nourished plant more succulent and more inviting than the inadequately nourished plant. In our monoculture practice, huge populations of corn plants per unit acre, well nourished, with a larger number of ears, provide the ideal target for the hungry insect. Therefore, insect infestation does not mean that the pesticides have failed; it is in large part the biological consequence of optimizing the target.

Does the individual farmer benefit by this application of technology, including pesticides? The average price of corn per bushel over the past 5 years has been $2.54 per acre, and therefore, for 1000 acres, the farmer's return was $1000 \times 2.54 \times 105.4$ = approximately $268,000. Insect losses at 12% can be approximately averaged at $32,200. Had the farmer employed the technology of 1941, his intake would have been $31.9 \times 1000 \times \$2.54$ = approximately $81,000. He would have had a 4% loss or $3,240.

The farmer's uncorrected gross income would total $268,000 − $32,200 = $235,800 with 1984 technology or $81,000 − $3,240 = $77,760 with 1941 technology. The difference between these numbers is the estimated gross income to the farmer who used 1984 technology (pesticides, fertilizer, seed varieties, etc.). His gross income increase is $235,800 − $77,760 = $158,040. In addition to his insecticide expenses, he used more fertilizer, fungicides, and other additives. If we regard this input increase as perhaps 20%, the corrected income increase is $0.80 \times \$156,000$ = $124,800 more than with 1941 technology.

Do the people of the United States benefit? For the country as a whole, the additional wealth created by technology can be crudely estimated. Because 73 million acres was involved in the corn harvest in 1982, the additional national wealth is $7.3 \times 10^4 \times 1.25 \times 10^5 = \$9,125,000,000!$

This number, which is a very rough approximation, is in itself not significant, but it is illustrative. Agriculture at the 1941 technological level would have been competitively impossible 20 or 40 years later. Modern technology, including pesticides, leads to extraordinary gains in productivity and wealth.

In fact, today the problem that faces U.S. agriculture is the difficulty of selling huge quantities of agricultural products. This new problem, which is outside the scope of this chapter, has resulted not from the failures of technology but as an unintended consequence of its success.

The objective of technology is to increase productivity. Insecticides continue to be an important part of that technology. Without them, in certain years crop yields would drop disastrously.

The problems that beset agriculture in the United States and in the rest of the world are social, political, and economic and would be severely exacerbated without modern technology. There is a worldwide need for food, and those who need it most cannot pay for it. Only science and technology, including pesticides, can provide the needed food for a burgeoning world population.

Special Problems of the Third World and Use of Agrochemicals

The drive for maximizing yields has affected pest control strategies in the economically advanced nations and certainly in much of the Third World, where higher yields are the difference between outright death by starvation and a low level of subsistence. Problems of environmental deterioration are recognized by the political and scientific leaders of the developing nations. However, they feel that the immediate requirements for subsistence today take precedence over the considerations of long-time trends. Their value system is certainly comprehensible.

Furthermore, the newer chemicals of the 1980s are of such complexity that whatever industry exists in the developing nations, private and government owned, cannot easily be adapted to manufacture these chemicals. Import is then necessary. Because these countries are characterized by an almost total lack of hard currency, only long-term loans, United Nations and World Bank intervention, or advanced country loans or gifts can provide them with the more modern compositions. No wonder many of the countries of Africa, Asia, and Latin America either import at low cost or manufacture some of the older and environmentally questionable pesticides.

Their use of pesticides that contaminate the environment is not a consequence of ignorance or stupidity; it is one additional result of grinding poverty.

A Reassessment of *Silent Spring*

Rereading *Silent Spring* in no way altered my original assessment that the volume is important and would affect the public's attitudes relative to the use of pesticides and their impact on the environment. Carson was a biologist and a consummate artist. Her earlier volume, *The Sea Around Us*, also demonstrated both of these capacities. *Silent Spring* preaches what Edith Efron calls an "apocalyptic vision" (4). Continuation of the large-scale production and use, particularly though not exclusively of the chlorinated hydrocarbon pesticides, DDT, the cyclodienes, 2,4-D, and other such pesticides would have had disastrous consequences on many nontarget organisms (birds, planktonic species, and certain large animals).

Not all Carson's predictions have come true. Birds do sing, and we are living longer than ever before. The legacy of Carson may be found in the legislation, some good and some not so good, and in the public belief, well justified, in the need to defend the environment against further deterioration.

Separated by a century in history, *Silent Spring* and Harriet Beecher Stowe's *Uncle Tom's Cabin* can be compared. Both books created social movements that have vitally affected American history. Another book with apocalyptic vision and far-reaching effects was George Orwell's *1984*. The public in the United States and Britain recognized the dangers described, and existing institutions in these countries and some others possessed the flexibility to enable them to avoid the horrors that were prophesied. Because of *Silent Spring*, the actions of industry, agriculture, and government have averted the worst consequences that were predicted.

It is necessary, however, both to recognize Carson's achievements and to discuss some of her inaccuracies. Examine the following quotations from *Silent Spring*:

> "With the advent of man the situation began to change, for man alone of all forms of life can create cancer-producing substances...." (p. 219)

> "Natural carcinogenic agents are few in number." (p. 220)

> "We are now aware of an alarming increase in malignant disease." (p. 121)

These generalizations are aberrations and misjudgements, and they should be focused upon because Carson was so successful in her influence on public policy. She was a prophet, but her present-day disciples are less discerning but more strident. They well may obstruct the future of both science and our nation's economic development. For example, as an extension of Carson's view of benign nature, it is "now believed... that at least 80% of all cancers are caused by synthetic chemicals with unfamiliar names like polyvinyl chloride, styrene oxide, urethane, diethylstilbestrol, and methylcholanthrene".

This perception now pervades all strata of U.S. society, influences the political process, and results in legislation, some of it harmful. A large proportion of our youth, the middle class, many of the affluent, and even members of the scientific community hold fast to the admixture of fear and faith expressed in the attitude that the results of the Industrial Revolution are detrimental to human health. However, this attitude has been refuted successfully (4–7). Carson was concerned with agrochemicals, but the movement she galvanized embraces a much wider scope.

The overwhelming statistics are, of course, life expectancies and mortality figures for cancer. We are creating a new social problem, but it is not the one that Carson's disciples envisaged. It is a demographic problem of a rapidly increasing aged population to be supported by a decreasing youthful population. Certain risks related to cancer and ill health are avoidable in the United States, chiefly tobacco smoking. Chemical manufacture makes a difficult-to-measure, but small, contribution to carcinogenesis of the total population. We have always had natural carcinogens. Trace quantities of pesticides in food make a relatively small but still unquantified contribution. Workers in chemical plants and professional chemists are at some risk because of their opportunities for high exposure.

Regulatory procedures for the manufacture of chemicals continue to be required. Most industrial scientists agree that reasonable regulation of agrochemicals continues to be necessary and desirable. However, prudence in administration of existing legislation need not require departure from scientific principles, and yet, many of the protocols used are scientifically questionable. The long-term feeding test at maximum tolerated dose is biochemically insupportable. Normal defense mechanisms are overloaded. Mammals, including humans, would have toxic or even lethal responses to sugar and salt fed at the maximum tolerated dose. Another questionable guideline is for absolute zero pesticide residues in food commodities for substances that show carcinogenic potential. Many essential minerals, vitamins, etc. would not pass this barrier. Finally, the ideas that no threshold dose exists for carcinogens and that all risk can be extrapolated linearly are questionable because they disregard essential substances that have mutagenic or carcinogenic properties.

Many current beliefs about cancer, industrial chemicals, genetic alteration, nuclear energy, food additives, natural foods, and even fluorides in drinking water are inaccurate, and their risks are grossly exaggerated. History records that such fears nearly halted vaccinations in the 19th century. Do we ban cars or do we try to educate and regulate their use, despite the fact that about 50,000 deaths occur each year? Should we ban penicillin (it caused tens of thousands of deaths and will cause very large numbers of deaths in the future) until we provide complete safety for the allergic individual?

Yes, there are risks with food additives, pesticides, and nuclear energy. Our science and technology does strive to minimize those risks, and our new technologies will continue to advance. Sensible regulation and reasonable controls are necessary parts of progress.

Consider again the two conflicting values: productivity and preservation of the environment. They cannot be taken alone or separately; they must be balanced and accommodated. The chemicals we now manufacture are better and less destructive. We need legislative improvements and administrative flexibility. On the whole, we are moving in these directions.

Glossary

Acute toxicity   Toxicity derived from a single application of a substance, usually occurring quickly, e.g., a few hours or days later.

Chlorinated hydrocarbons   Compounds that contain relatively large proportions of chlorine, carbon, and hydrogen. Occasionally other elements are included, such as oxygen. These chemicals are usually persistent in the environment and accumulative in biological systems.

Chronic toxicity   Toxicity resulting from continuous periodic application of a substance, usually occurring weeks, months, or years later.

Cyclodienes   Compounds derived from the molecule hexachlorocyclopentadiene, a ring compound made up of chlorine and carbon.

Enzymatically vulnerable compounds   Those that are transformed by *enzymes* (natural catalysts) derived from living organisms.

Organic chemical pesticides   Those derived from carbon compounds. Earlier chemical pesticides were generally *inorganic* (containing no carbon), e.g., sulfur, lead arsenate, and copper sulfates.

Photochemically vulnerable chemicals   Those that are transformed by light into other substances.

Pyrethrum   A natural product derived from certain plants that possesses insecticidal properties. *Pyrethroids* are synthetic organic compounds with some resemblances in structure and activity to the natural product. The pyrethroids are generally more chemically stable than the natural product and are more biologically potent.

Acknowledgments

I gratefully acknowledge Noreen Collins, Christina Darle, Christine Miller, and Luana Staiger of Zoecon Corporation for helping prepare the typescript and the slides; and Ted Eichers of the U.S. Department of Agriculture for sending me much statistical data.

A disclaimer is necessary. The judgements that were made on all extra-scientific matters are my own. They do not represent any official views of my current employer (Zoecon Corporation), Chevron Chemical Company (my erstwhile employer), the National Agricultural Chemicals Association, or any other group.

Literature Cited

1. Melnikov, N. N. *Residue Rev.* **1971,** *36,* 5.
2. Swift, Jonathan "Voyage to Brobdingnag", *Gulliver's Travels;* Oxford University Press: New York, 1977.
3. Woodwell, G. M.; Craig, P. P.; Johnson, H. A. *Science* **1971,** *174,* 1101.
4. Efron, E. *The Apocalyptics;* Simon & Schuster: New York, 1984.
5. Doll, R.; Peto, R. *The Cause of Cancer;* Oxford University Press: New York, 1981.
6. Ames, B. *Science* **1983,** *221,* 1256.
7. Epstein, S. S.; Swartz, V. B.; Ames, B. *Science* **1984,** *224,* 658, et. seq.

Many statistics and other data given in this chapter were taken from the following references:

• *Agricultural Statistics;* U.S. Government Printing Office: Washington, DC. A yearly publication with some data in each volume that compares the items studied with past years' performance.

- "The Pesticide Industry", by Gustave K. Kohn, in *Riegel's Handbook of Industrial Chemistry*, 8th ed.; Kent, James, A., Ed.; Van Nostrand Reinhold: New York, 1983.

- *Quantities of Pesticides Used by Farmers in 1966*; Agricultural Report No. 179, Environmental Research Service, U.S. Department of Agriculture, U.S. Government Printing Office: Washington, DC.

- *Outlook and Situation (Pesticides)*; IOS-2, Environmental Research Service, U.S. Department of Agriculture, U.S. Government Printing Office: Washington, DC, 1983.

11❧Is Silent Spring *Behind Us?*

David Pimentel

Is *Silent Spring* behind us? Have environmental problems associated with pesticide use improved? The answer is a qualified "yes".

Rachel Carson's warning in 1962 generated widespread concern, but many years elapsed before action was taken to halt some of the environmental damage being inflicted by pesticides on our sensitive natural biota. More than 20 years later we still have not solved all the pesticide environmental problems, although some real progress has been made.

Fewer Pesticide Problems
During the Past Two Decades

Chlorinated insecticides, such as DDT, dieldrin, and toxaphene, are characterized by their spread and persistence in the environment. The widespread use of chlorinated insecticides from 1945 to 1972 significantly reduced the populations of predatory birds such as eagles, peregrine falcons, and ospreys (1). Trout, salmon, and other fish populations were seriously reduced, and their flesh was contaminated with pesticide residues. Snakes and other reptile populations, as well as certain insect and other invertebrate populations that were highly sensitive to the chlorinated insecticides (1) were reduced.

Since the restriction on the use of chlorinated insecticides went into effect in 1972, the quantities of these residues in

0980-4/87/0175$06.00/0 © 1987 American Chemical Society

humans and in terrestrial and aquatic ecosystems have slowly declined. From 1970 to 1974, for example, DDT residues in human adipose tissue declined by about one-half in Caucasians who were 0–14 years of age (2) (see Table 1). The declines in other Caucasian age groups and in blacks have not been as great. In agricultural soils, DDT residues have declined by about one-half or from 0.015 parts per million (ppm) in 1968 to 0.007 ppm in 1973 (3). The decline of DDT in soil led to a decline in the amount of DDT running into aquatic ecosystems and resulted in a significant decline in DDT residues found in various fish. For example, in lake trout caught in the Canadian waters of eastern Lake Superior, DDT residues declined from 1.04 ppm in 1971 to only 0.05 ppm in 1975 (4). In aquatic birds that feed on fish, DDT residues also declined. For example, DDT residues in brown pelican eggs collected in South Carolina declined from 0.45 ppm in 1968 to only 0.004 ppm in 1975 (5).

Table 1. **Total DDT Equivalent Residues in Human Adipose Tissue from General U.S. Population by Race**

Age (years)	1970	1971	1972	1973	1974
Caucasians					
0–14	4.16	3.32	2.79	2.59	2.15
15–44	6.89	6.56	6.01	5.71	4.91
45 and above	8.01	7.50	7.00	6.63	6.55
Negroes					
0–14	5.54	7.30		4.68	3.16
15–44	10.88	13.92	11.32	9.97	9.18
45 and above	16.56	19.57	15.91	14.11	11.91

NOTE: All residues are measured in parts per million lipid weight.
SOURCE: Reference 2.

Because DDT and other organochlorine residues in terrestrial ecosystems have declined, various populations of birds, mammals, fishes, and reptiles have started to recover and increase in number. For example, peregrine falcons have been bred in the laboratory and then successfully released in the environment. Limited data do exist on the recoveries of a few animal species, but we do not know the recovery rates for those animal populations that were seriously affected by chlorinated insecticides. Those species with short generation times and high reproductive rates, like insects, have probably recovered best.

New pesticide regulations established in the early 1970s restricted the use of highly persistent pesticides, which include chlorinated insecticides. DDT, toxaphene, and dieldrin, for example, persist in the environment for 10 to 30 years (1). Two major problems are associated with the use of highly persistent pesticides. Annual applications of chlorinated insecticides add to the total quantity of insecticides in the environment because they degrade slowly. This persistence in the environment increases the chances for the chemicals to move out of the target area into the surrounding environment.

The amount of chlorinated insecticide residues in the environment since most of the chlorinated insecticides were banned has been declining. But because these insecticides are relatively stable, some will persist 30 years or more, and some will be present in the U.S. environment until the end of this century. Fortunately these residues are relatively low, so their effect on most organisms should be minimal.

Persistence of chlorinated insecticides in the environment is only one of the problems created by these chemicals. Their solubility in fats and oils resulted in their accumulation in the fatty tissues of animals, including humans (6). Thus, bioaccumulation of chlorinated insecticides is a serious environmental problem. Organisms like water fleas and fish, for example, concentrated DDT and other chlorinated insecticides from a dosage of 1 part per billion (ppb) in the environment to levels in their tissues of 100,000 times that (1). Bioaccumulation continues in the environment with several pesticides (e.g., parathion and 2,4-D), but restricting the use of chlorinated insecticides has reduced this environmental problem.

Movement and magnification of pesticides in the food chain also occurs, but must be carefully documented (6). Some organisms concentrate pesticides in their bodies 100,000-fold over levels in the ambient environment, and this condition might mistakenly be interpreted as a case of biomagnification in the food chain. Biomagnification in the food chain has been documented with birds like osprey and gulls that feed on fish and has proven to be a serious problem to these predaceous birds (1).

Increased Pesticide Problems
During the Past Two Decades

Although restricted use of chlorinated insecticides has relieved some environmental problems, the escalation of pesticide use since 1970 has intensified several other environmental and social problems. Pesticide production and use has increased 2.3-fold since 1970, from around 1.0 to nearly 1.5 billion pounds annually (7).

Recent research has documented the fact that certain pesticide use may actually increase pest problems. For example, herbicides like 2,4-D used at recommended dosages on corn increased the susceptibility of corn to both insects and plant pathogens (8). Also the reproduction of certain insects can be stimulated by low dosages of certain insecticides, as occurred in the Colorado potato beetle. For example, sublethal doses of parathion increase egg production by 65%. In addition, most of the insecticides that replaced the chlorinated insecticides are more toxic per unit weight than the chlorinated insecticides.

If one pesticide is more toxic and more biologically active than another, it is not necessarily hazardous to the environment. Risk depends on the dosage and method of application of the specific pesticide. If one pesticide's per-unit weight is more toxic than another, the more toxic chemical is usually applied at a lower dosage that will cause about a 90% kill in the pest population. Thus, a highly toxic material used at a low dosage can achieve about the same mortality as a low-toxicity material. Both high- and low-toxicity pesticides affect pests and nontarget organisms in a similar manner, but the risks to humans handling highly toxic pesticides are far greater than when handling pesticides with a low toxicity. Humans handling highly toxic pesticides like parathion are more likely to be poisoned than those handling pesticides of low toxicity like DDT. If one spills DDT and wipes the pesticide off the skin, no harm is done. However, a similar accident with parathion often leads to poisoning severe enough to require hospitalization.

Human Poisonings

Humans are exposed to pesticides by handling and applying them, by contacting them on treated vegetation, and, to a lesser extent, from their presence in food and water supplies. The number of annual human pesticide poisonings has been estimated at about 45,000; about 3000 of these are sufficiently severe to require hospitalization (*9, 10*). The number of annual accidental deaths caused by pesticides is about 50. Accurate data on human pesticide poisonings still are not available 20 years after *Silent Spring*.

Furthermore, detecting the causes of cancer from pesticides is exceedingly difficult because of the long lag time prior to illness and the wide variety of cancer-producing factors that humans are exposed to in their daily activities. No one knows if less human cancer is caused by pesticides now than 20 years ago. Probably less than 1% of human cancers today are caused by pesticides (*11*).

We are constantly exposed to pesticides. Despite efforts to keep pesticides out of our food and water, about 50% of U.S. foods sampled by the Food and Drug Administration (FDA) contain detectable levels of pesticides (*12–14*). Improvements in analytical chemical procedures are helping us detect smaller and smaller quantities of pesticides in food and water. These extremely low dosages should have little or no public health effect.

Domestic Animal Poisonings

Because domestic animals are present on farms and near homes where pesticides are used, many of these animals are poisoned. Dogs and cats are most frequently affected because they often wander freely about the home and farm and have ample opportunity to come in contact with pesticides (*10*).

A major loss of livestock products (about $3 million annually) occurs when pesticide residues are found in these products (*10*). This problem will probably continue as the quantity of pesticides used continues to rise.

Bee Poisonings

Honeybees and wild bees are essential to the pollination of fruits, vegetables, forage crops, and natural plants (*15*). Pesticides kill bees, and the losses to agriculture from bee kills and the related reduction of pollination are estimated to be $135 million each year (10). Evidence suggests that bee poisonings are probably greater now than in 1962 for several reasons. More highly toxic insecticides are being used, and greater quantities of insecticides are being dispensed. In addition, more pesticide is being applied by aircraft, and aircraft applications are employing ultra low volume (ULV) application equipment. ULV applications require smaller droplets for coverage, and this practice tends to increase pesticide drift problems.

Crop Losses

Although pesticides are employed to protect crops from pests, some crops are damaged as a result of pesticidal treatments. Heavy pesticide use damages crops and causes declines in yields because: (1) herbicide residues that remain in the soil after use on one crop injure chemically sensitive crops planted in rotation, (2) certain desired crops cannot be planted in rotation because of knowledge of potential hazard injury, (3) excessive residues of pesticides remain on the harvested crop and result in its destruction or devaluation, (4) pesticides that are applied improperly or under unfavorable environmental conditions result in drift and other problems, and (5) pesticides drift from a treated crop to nearby crops and destroy natural enemies or the crop itself.

Although an accurate estimate of the negative impact of pesticides on crops in agriculture is extremely difficult to obtain, a conservative estimate is about $70 million annually (10). The problem is probably worse today than in Carson's time because 7 times more pesticide is being applied today than 20 years ago, and its use is more widespread. This statement is especially true of herbicides.

Reduced Populations of Natural Enemies

In undisturbed environments, most insect and mite populations remain at low densities because a wide array of factors, including natural enemies, control them (16–18). When insecticides or other pesticides are applied to crops to control one or more pest species, natural enemy populations are sometimes destroyed, and subsequently pest outbreaks occur (6, 19).

For example, before the synthetic pesticide era (1945) the major pests of cotton in the United States were the boll weevil and cotton leafworm (*20*). When extensive insecticide use began in 1945, several other insect and mite species became serious pests. These include the cotton bollworm, tobacco budworm, looper, cotton aphid, and spider mites (*20*). In some regions where pesticides are used to control the boll weevil, as many as five additional treatments have to be made to control boll-worms and budworms because their natural enemies have been destroyed (*21*). This cycle has meant more pesticide use, more natural enemies destroyed, greater pest populations, and more pesticides used.

Pesticide Resistance

In addition to destroying natural enemies, the widespread use of pesticides often causes pest populations to develop resistance and pass it on to their progeny. More than 420 species of insects and mites and several weed species have developed resistance to pesticides (*22*). Pesticide resistance in pests results in additional sprays of some pesticides or the use of alternative and often more expensive pesticides. Again the process of pest control escalates the cycle of pesticide use and the development of resistance.

An estimated $133 million worth of added sprays or more expensive pesticides has been employed to deal with the resistance problem annually (*10*). This dollar cost, of course, does not include the side effects apparent in the environment and in public health from using more pesticides and more toxic pesticides.

Fishery Losses

Pesticides in treated cropland often run off and move into aquatic ecosystems. Water-soluble pesticides are easily washed into streams and lakes, whereas other pesticides are carried with soil sediments into aquatic ecosystems. Each year several million tons of soil, and with it, pesticides, are washed into streams and lakes annually (*23, 24*).

At present only a small percentage of fish kills are reported because of the procedures used in reporting fish losses (25). For example, 20% of the reported fish kills give no estimate of the number of dead fish because fish kills often cannot be investigated quickly enough to determine the primary cause. Also, fast-moving waters rapidly dilute all pollutants, including pesticides, and thus make the cause of the kill difficult to determine. Dead fish are washed away or sink to the bottom, so accurate counts are not possible.

Samples of water recently confirmed a steadily decreasing concentration of pesticides found in surface waters and streams from 1964 to 1978 (26–28). This reduction is apparently related to the replacement of persistent pesticides with less persistent materials. Despite the reduced pesticide residues in streams, an estimated $800,000 or more in fish is lost annually (each fish was calculated to have a value of 40 cents) (10). This estimate of nearly $1 million probably is several times too low and does not confirm that *Silent Spring* is behind us.

Impacts on Wildlife and Microorganisms

Too little information exists to make even a conservative estimate of the populations of vertebrates, invertebrates, and microorganisms that are adversely affected by pesticides (10). Most invertebrates and microorganisms perform many essential functions to agriculture, forestry, and other segments of human society; such as preventing the accumulation of water, cleaning water or soil of pollutants, recycling vital chemical elements within the ecosystem, and conserving soil and water (10). An estimated 200,000 species of plants and animals exist in the United States and, at best, we have information on the effects of pesticides on less than 1000 species. Most of these data are based on "safe concentration" tests conducted in the laboratory. This situation confirms that little is known about pesticide effects on the natural environment. At present evaluation must be based on indicator species.

Status of Integrated Pest Management

Integrated pest management (IPM), introduced more than a decade ago, aimed to reduce pesticide use by monitoring pest populations and using pesticides only when necessary as well as augmenting pest control with alternative nonchemical strategies (*29*). What happened? IPM has not been successful (*30*), and in fact, more of all kinds of pesticides are being used in the United States and throughout the world than ever before.

The reasons for the poor performance of IPM are complex. First, IPM technology, even if it is simply monitoring pest and natural enemy populations, requires a great deal more basic information than scientists now have. This fact signals the pressing need for basic research on the ecology of pests, their natural enemies, and their environment. Also, the use of this basic information to develop control programs is much more sophisticated than routine application of pesticides. Because this technology is more sophisticated, trained manpower is needed, and often the farmer is not trained and cannot be expected to carry out effective IPM programs.

Pesticides are unquestionably simple and quick to use. They have a significant psychological advantage over IPM and especially over nonchemical controls like biological control. Biological controls gradually bring pest populations under control, but do not give the immediate satisfaction of direct kill like pesticides do. However, as research continues and greater ecological knowledge of pests and agroecosystems increases, IPM has the potential to improve pest control.

Why Are Losses Due to Pests Greater Today Than 40 Years Ago?

Currently, an estimated 37% of all crops is lost annually to pests (13% to insects, 12% to plant pathogens, and 12% to weeds) in spite of the combined use of pesticidal and nonchemical controls (*31*). According to a survey of data collected from 1942 to the present, crop losses from weeds declined slightly from 13.8% to 12% because of a combination of improved herbicidal, mechanical, and cultural weed control practices. During the same period, losses from plant pathogens increased slightly from 10.5% to 12% (*32*).

On average, however, crop losses due to insects have increased nearly twofold (from 7% to about 13%) from the 1940s to the present (*32*) in spite of a 10-fold increase in insecticide use. Thus far the impact of this loss in terms of production has been effectively offset through the use of higher yielding varieties and increased use of fertilizers (33).

The substantial increase in crop losses caused by insects can be accounted for by some of the major changes that have taken place in U.S. agriculture since the 1940s. These changes include

- planting of crop varieties that are increasingly susceptible to insect pests;

- destruction of natural enemies of certain pests, which in turn creates the need for additional pesticide treatments (19);

- increase in the development of pesticide resistance in insects (34–36);

- reduced crop rotations and crop diversity and an increase in the continuous culture of a single crop (18, 21, 37);

- reduced FDA tolerance and increased cosmetic standards of processors and retailers for fruits and vegetables (38);

- reduced field sanitation including less destruction of infected fruit and crop residues;

- reduced tillage, leaving more crop remains on the land surface to harbor pests for subsequent crops (39);

- culturing crops in climatic regions where they are more susceptible to insect attack (37);

- use of pesticides that alter the physiology of crop plants and make them more susceptible to insect attack (8, 40).

Conclusion

Progress has been made on pesticide problems, but *Silent Spring* is not entirely behind us. Pesticide use continues, and the quantities of pesticides applied grow annually despite support for IPM control. In future decades, as the world population grows rapidly and agricultural production is stretched to meet food needs, we should not forget Carson's warnings.

Pesticides will continue to be effective pest controls, but the challenge now is to find ways to use them judiciously to avoid many of the environmental hazards and human poisonings that exist today. With this goal for research and development we can achieve effective, relatively safe pest control programs.

Glossary

Biomagnification   The escalating concentration of pesticides from soil, water, and air in plants and animals and, in turn, in the food chains of biological systems.

Pathogens   Microorganisms that cause severe disease in plants and animals.

Literature Cited

1. Pimentel, D. *Ecological Effects of Pesticides on Nontarget Species;* U.S. Government Printing Office: Washington, DC, 1971.
2. Kutz, F. W.; Yobs, A. R.; Strassman, S. C.; Viar, J. F. *Pestic. Monit. J.* **1977,** *11,* 61–63.
3. Carey, A. E. *Pestic. Monit. J.* **1979,** *13,* 23–27.
4. Frank, R.; Holdrinet, M.; Braun, H. E.; Dodge, D. P.; Sprangler, G. E. *Pestic. Monit. J.* **1978,** *12,* 60–68.
5. Blus, L. J.; Lamont, T. G.; Neely, B. S. *Pestic. Monit. J.* **1979,** *12,* 172–184.
6. Pimentel, D.; Edwards, C. A. *Bioscience* **1982,** *32,* 595–600.
7. *Statistical Abstract of the United States 1985.* 105th ed. U. S. Bureau of the Census: Washington, DC, 1984.
8. Oka, I. N.; Pimentel, D. *Science* **1976,** *193,* 239–240.
9. *National Study of Hospital-Admitted Pesticide Poisonings;* Environmental Protection Agency. Epidemiologic Studies Program. Human Effects Monitoring Branch. Technical Services Division. Office of Pesticide Programs. U.S. Government Printing Office: Washington, DC, April 1976.
10. Pimentel, D.; Andow, D.; Dyson-Hudson, R.; Gallahan, D.; Jacobson, S.; Irish, S.; Moss, A.; Schreiner, I.; Shepard, M.; Thompson, T.; Vinzant, B. *Oikos* **1980,** *34,* 127–140.
11. Schotterfeld, D., personal communication, Sloan–Kettering Cancer Center, 1978.
12. Johnson, R. D.; Manske, D. D. *Pestic. Monit. J.* **1977,** *11,* 116–131.
13. McEwen, F. L.; Stephenson, G. R. *The Use and Significance of Pesticides in the Environment;* Wiley: New York, 1979.

14. Johnson, R. D.; Manske, D. D.; New, D. H.; Podrebarac, D. S. *Pestic. Monit. J.* **1981,** *15,* 39–50.
15. McGregor, S. E. *Insect Pollination of Cultivated Crop Plants;* Agricultural Handbook No. 496. U.S. Department of Agriculture. Agricultural Research Service. U.S. Government Printing Office: Washington, DC, 1976.
16. *Biological Control of Insect Pests and Weeds;* DeBach, P. H., Ed.; Reinhold: New York, 1964.
17. *New Technology of Pest Control;* Huffaker, C. B., Ed.; Wiley: New York, 1980.
18. Pimentel, D. *Ann. Entomol. Soc. Am.* **1961,** *54,* 76–86.
19. van den Bosch, R.; Messenger, P. S. *Biological Control;* InText Educational Publishers: New York, 1973.
20. Newsom, L. D. *Proc. Boll Weevil Research Symp.* State College, MS, 1962; pp 83–94.
21. Pimentel, D.; Shoemaker, C.; LaDue, E. L.; Rovinsky, R. B.; Russell, N. P. *Alternatives for Reducing Insecticides on Cotton and Corn: Economic and Environmental Impact;* Report on Grant No. R802518-02. Environmental Protection Agency. U.S. Government Printing Office: Washington, DC, 1977.
22. *Pest Resistance to Pesticides;* Georghiou, G. P.; Saito, T., Eds.; Plenum: New York, 1983.
23. *Impacts of Technology on Productivity of the Croplands and Rangelands of the United States;* Office of Technology Assessment. U.S. Government Printing Office: Washington, DC, 1982.
24. *Agriculture's Soil Conservation Programs Miss Full Potential in the Fight Against Soil Erosion;* U.S. General Accounting Office. U. S. Government Printing Office: Washington, DC, 1983.
25. *Fish Kills Caused by Pollution in 1970–1974;* Environmental Protection Agency. U.S. Government Printing Office: Washington, DC, 1972–1976.
26. Lichtenberg, J. J.; Eichelberger, J. W.; Dressman, R. C.; Longbottom, J. E. *Pestic. Monit. J.* **1970,** *4,* 485–488.
27. Schulze, J. A.; Manigold, D. B.; Andrews, F. L. *Pestic. Monit. J.* **1973,** *7 (1),* 73–84.
28. *Monthly Report on Toxic Substances Impacting on Fish and Wildlife;* Fish and Wildlife Division. New York State Department of Environmental Conservation. Apr.–Feb. 1977–1978.
29. Pimentel, D. *Crop Protect.* **1982,** *1,* 5–26.
30. "Pesticides, Food, Health, and Environment: Problems and Needs"; United Nations Environmental Programme. Draft manuscript from meeting on Environmental Effects of Chemicals, Geneva, Feb. 6–10, 1984.
31. *Handbook of Pest Management in Agriculture;* Pimentel, D., Ed.; CRC: Boca Raton, FL, 1981; Vols 1–3.
32. Pimentel, D.; Krummel, J.; Gallahan, D.; Hough, J.; Merrill, A.; Schreiner, I.; Vittum, P.; Koziol, F.; Back, E.; Yen, D.; Fiance, S. *Bioscience* **1978,** *28,* 772, 778–784.

33. Pimentel, D.; Hurd, L. E.; Bellotti, A. C.; Forster, M. J.; Oka, I. N.; Sholes, O. D.; Whitman, R. J. *Science* **1973**, *182*, 443–449.
34. Georghiou, G. P. *Annu. Rev. Entomol.* **1972**, *3*, 122–168.
35. Hance, R. J. Presented at the Southeast Asian Workshop on Pesticide Management, Bangkok, Thailand, 1977.
36. Pimentel, D.; Goodman, N. In *Survival in Toxic Environments;* Khan, M. A. Q.; Bederka, J. P., Eds.; Academic: New York, 1974; pp 25–52.
37. Pimentel, D. In *Origins of Pest Parasite, Disease, and Weed Problems;* Cherrett, J. M.; Sagar, G. R., Eds.; Blackwell: Oxford, 1977; pp 3–31.
38. Pimentel, D.; Terhune, E. C.; Dritschilo, W.; Gallahan, D.; Kinner, N.; Nafus, D.; Peterson, R.; Zareh, N.; Misiti, J.; Haber-Schaim, O. *Bioscience* **1977**, *27*, 178–185.
39. Musick, G. J.; Petty, H. B. In *Conservation Tillage;* Soil Conservation Society of America: Ankeny, Iowa, 1973; pp 120–125.
40. Dunham, E. W.; Clark, J. C. *J. Econ. Entomol.* **1941**, *34*, 587–588.

In Conclusion

Rachel Carson in 1961 in Maine.

(Photograph by Erich Hartmann. Used by permission of the Rachel Carson Council, Inc.)

12 Many Roads and Other Worlds

*Gino J. Marco, Robert M. Hollingworth,
and William Durham*

Rachel Carson concluded *Silent Spring* with a chapter entitled, "The Other Road", proposing a biological, essentially nonchemical approach to pest control. In summarizing this revisit, we examine her ideas with the

0980–4/87/0191$06.00/0 © 1987 American Chemical Society

perspective of almost 25 years of experience to ask whether she was right, where the issues she addressed now stand, and what type of controls on usage and safety have evolved. In one respect at least, our book is likely to fail: we cannot match Rachel Carson's language and imagery. These chapters were written by practicing scientists, schooled in the need to be conservative in their interpretation of data, to write directly, and to avoid emotionalism.

Chemicals To Control Pests

The control of pests has been desired throughout more than 10,000 years of farming because pests have always competed for food. As early as 3000 years ago, sulfur was used; as time progressed, copper, lead, and arsenic compounds were added. These compounds have had some human health effects, but the environmental consequences were unrecognized. Thus, simple, generally inorganic chemicals have been used for a long time. However, the introduction of synthetic organic chemicals as a new technology, especially the organochlorine pesticides typified by DDT, led to the issues expressed in *Silent Spring*.

Risk Versus Benefit

In the 19th Century, 90% of the population in the United States was needed to supply food and fiber, compared with about 4% today. The advances in technology that have allowed this efficiency in modern agriculture have included much use of chemicals. Tillage practices and farm labor could be reduced, while farm production continually increased. These are the visible benefits of pesticide use.

The overuse and misuse of persistent organochlorine insecticides were Carson's main concerns. She felt that persistent organochlorine insecticides and some other pesticides abused the environment, were hazards for wildlife, and possibly even caused toxic effects in humans. These are the hidden costs of pesticide use.

Silent Spring led society to evaluate the new technologies in terms of risk versus benefits rather than on the basis of benefits alone. In fact, many people feel that risk has become the predominant concern. As the public's concern grew from apathy to advocacy, Congress responded with a series of amendments to already existing laws, mandating the protection of public health and the environment. This emphasis was catalyzed by *Silent Spring* and continues in the present.

Analysis of Pesticides

To establish risk and define safe usage, the amount of pesticide present in a sample must be assessed. In Carson's day, analyses were done mainly by using intensity of color to measure the amounts present. These procedures required large samples, were time-consuming, were not highly sensitive (i.e., the state of the art was in the range of parts per million), and they were generally not selective (i.e., they measured more than one chemical, but often interpreted the measurement as only one). Even the later introduction of more sensitive and specific analytical methods, such as gas chromatography, did not lead to unequivocal results; DDT values from the 1960s may be unreliable because DDT was confused with other organochlorine compounds such as polychlorinated biphenyls (PCBs).

Silent Spring served to encourage funding for improved analytical procedures and instrumentation. These concerted efforts, along with added regulations, have brought about an improved scientific base of information upon which decisions about safety are based. Since *Silent Spring*, analytical chemical techniques have greatly increased sensitivity in the detection of residues, in some cases, such as dioxin, by at least one million times. Furthermore, acceptable analyses usually measure only the particular chemical desired.

Evaluating Risk

Whereas lower levels can now be detected, the true significance of these small amounts of pesticide residues is often quite uncertain. When a measurable but very small amount of a chemical is present in the environment or food supply, perspective on its significance is often lost. All chemicals can be toxic at a sufficiently high level. The risk occurs when the chemical exceeds its toxic level, which may be considerably above the lower limit for detectability in the environment.

Although amounts present can be determined with high

194 / SILENT SPRING REVISITED

accuracy, evaluating risk also requires the evaluation of the toxicity of the chemical. Toxicity evaluation is a complex process with great uncertainties, and it does not provide answers that are nearly as precise as the analytical techniques. Thus, scientists and regulators, in trying to relate low levels of exposure to their possible toxic effects in humans or other living systems, are forced to make intuitive value judgments that inevitably lead to controversies, even among themselves. The public often overestimates the ability of scientists to predict adverse effects. Risk is assumed avoidable, and an abundant food supply is taken for granted.

Persistence of Pesticides

One of the assumptions that *Silent Spring* challenged was that our environment, especially water, had an almost infinite capacity to cleanse itself. In fact, we have since learned to our surprise that highly persistent compounds can exist for many years in the environment, be transported globally, and accumulate in living organisms far from their site of application.

The residues in wildlife of the older organochlorine pesticides (e.g., DDT, aldrin, and dieldrin) have declined slowly but significantly since *Silent Spring*, because most are now limited in use or even banned. One of the more important legacies of *Silent Spring* was to stimulate an organized approach to studying chemical effects on our ecosystems. Such studies have greatly increased our knowledge of the fate of chemicals in the environment, yet new problems still arise.

Wells provide much of our drinking water, and so concerns over ground water contamination are becoming prevalent. The application of pesticides to broad expanses of farmland can contribute chemicals to ground water under a limited range of conditions, but such contamination seems more likely to come from hazardous waste disposal sites. Rachel Carson was prescient in alluding to such problems from chemical dumps. Also, the older persistent, fat-soluble, nonleaching pesticides have been replaced with chemicals that are relatively nonpersistent but more mobile in the soil. What often happens with advancing technology is that the solution of one problem generates a set of more complex problems.

Effects of Pesticides

Birds were the prime example used by Rachel Carson to illustrate and speculate upon the effects of pesticides on wildlife. The ban on DDT and related pesticides has decreased many undesired effects in birds, although some of the newer, more toxic insecticides have caused local problems in bird populations. Increased ecological information on chemicals enhances our understanding of such undesired effects. Modern strategies of pesticide development, including tests for toxicity in birds, are leading to products that minimize damage to bird populations. In addition, the public's desire to protect the environment has resulted in the modification and extension of regulations to protect wildlife.

Rachel Carson's concerns about the health effects of pesticides have led to a carefully managed approach in pesticide development covering chemical design, controlled formulation and use, and proper disposal, all with human safety in mind. With the use of pesticides for suicide and a sometimes casual or erroneous reporting of death, the extent of accidental poisonings is difficult to assess even in the United States. Also, the incidence of nonlethal injury to those using pesticides and possible long-term chronic effects are poorly defined and more controversial. Regulations to control worker exposure are continually revised and broadened; however, this safety net still has unavoidable weaknesses.

The long-delayed response and need to use animal models of debatable significance in toxicity testing has made cancer risk measurement for humans difficult, and in many cases, inexact. Thus, confusion still exists about the carcinogenic potential of many pesticides, and some older compounds have not been evaluated with methods deemed satisfactory by today's standards. However, Carson's view that "man alone of all life forms, can create cancer-producing substances" was seriously misleading because nature does indeed produce many such substances, for example, in the "bracken that once lifted high its proud lacework" and in the aflatoxins of *Aspergillus* mold, to name just two. This observation is not an argument to ignore the additional burden of synthetic compounds, but is meant simply to point out that powerful chemical poisons are commonly present in nature.

Response of Various Sectors

Industry has responded to the public's concerns by developing less-persistent pesticides that are not bioaccumulating, are applied at low rates, and have potential for use in integrated pest management (IPM) programs, that is, approaches that use many control methods and not chemicals alone. The government responded with new legislation to control all aspects of pesticide use, including health and environmental protection. The government continues this process to the present. Environmentalists responded by organizing advocacy groups. As a consequence, our society has not been allowed to forget or ignore the abuses described in *Silent Spring*. These problems are now less prevalent in the United States, and corrective action is generally quickly initiated once problem areas are defined. It is also becoming apparent that the broad utility of biological controls is still in the future. The often-proposed combination of biological and chemical controls will require many more technologists than are presently being trained.

Different Sides of the Coin

The diverse opinions and opposing views between manufacturer–user and environmentalist are clearly seen in the chapters of this book. In fact, the same information can be viewed in different ways. Our present high agricultural productivity is dependent on the use of monoculture techniques with great control of nutrients and pests, limited options for crop rotation, increasing biological resistance to existing pesticides, and the genetic selection of plants for high yields that sometimes ignores natural resistance to pests and disease.

The manufacturer–user interprets these situations as (1) a need for better-growing high-yield plants that can lead to increased insect infestation requiring more control; (2) a requirement for new, safe pesticides with different mechanisms of action; (3) a use of fertilization to produce lush plants, although possibly leading to more weed problems; and (4) a chance to increase the competitiveness and profitability of our agriculture in providing cheap and plentiful food of the highest quality. The environmentalist looks upon this high agricultural productivity as (1) a broader use of more pest-susceptible plants in inherently unstable monocultures, leading to more crop losses; (2) newer classes of pesticides with unpredictable side effects introduced into the environment; (3) a degradation of the soil integrity by overfertilization and overproduction at a time of agricultural surplus; and (4) a high-density farming strategy with heavy agrochemical inputs, resulting in more hidden costs and delaying the application of biological and other nonchemical controls.

The manufacturer–user looks on the described situations as opportunities to increase food and fiber for an ever-growing population. The environmentalist looks on the same picture as an increased assault on nature and human health, with the worry that Rachel Carson's warning will be forgotten.

Evaluation of Hazards

In developed countries that have control of disease and an adequate food supply, evaluation of the hazards of pesticides results primarily in concerns about long-term, low-level residues and environmental effects. In underdeveloped nations, the first priority is survival in the face of a tropical climate that maintains pest and disease pressures all year, where illiteracy and extreme poverty are present, and where few people are technically trained. These countries do not have the money to purchase or the resources to manufacture the newer, improved pesticides and must use the older, cheaper pesticides with their familiar problems.

Many of the issues discussed in *Silent Spring* now apply to these underdeveloped nations. Poor storage conditions, overuse, misuse, uncontrolled disposal, and excessive worker exposure are routine in these societies. Developed countries should take notice because food is exported from these nations, and global transportation of residues from overused pesticides could occur. After expressing such concerns to the underdeveloped nations, corrective action has often been taken, but overreaction and fear, which can be counterproductive, have also been noted.

In Summation

So where are we some 25 years after *Silent Spring*? With fewer farmers and less labor available, yet more mouths to feed, there seems no likelihood of completely eliminating chemical controls in the near future. The introduction of biological controls is far from a panacea and requires a more sophisticated approach involving many trained technologists. The new biotechnology of gene manipulation to produce, for example, pest-resisting plants appears to be a slow and costly process. The promise is great, but the eventual effect is unknowable.

Meanwhile, nature does not stand still in its ability to adapt to any challenge. We can expect pests to adapt to effective biological controls and engineered plants just as they have to chemicals. Furthermore, it is not certain in the long run that biological and biotechnological approaches to pest control will be safer than current chemicals to mankind and our environment. The other underdeveloped societies will most likely be brought into the domain of modern agriculture. They will leave behind the problem chemicals and train their people in the proper use of pesticides. And *Silent Spring* will still be around to refresh their memories.

Was Rachel Carson right? In many respects, yes. In her time, the environment was relentlessly assaulted by a society hoping for total control. Nature was not as self-cleansing as we believed. Many of Carson's predictions about environmental toxicity, human health effects, water contamination, and waste site problems have proved correct.

Was Rachel Carson wrong? In fewer respects, yes. Nature, not just humans, generates its share of carcinogens and other poisons. Nature and humans both use chemicals to their own advantage. The human life span is still increasing. Society has responded, and, of course, birds still sing. Biological controls alone have been able to replace chemicals only in a few specific cases (e.g., the screw worm and cottony cushion scale) (1). However, biological–chemical combinations, as in integrated pest management (IPM), have met with some success, mainly in controlling insects. Other pesticide use areas (such as diseases, weeds, and nematodes) are being investigated (2). Present approaches to IPM are to optimize the use of, not eliminate, chemicals. Rachel Carson set in motion a philosophy to use all tools in controlling pests, not to rely exclusively on chemicals.

Virtually all the issues Rachel Carson explored are in some stage of correction, and an aware public is now on guard. The future has many roads, not one. Some of these roads are as unexplored and unknown as that for the organic pesticides when they were first introduced. The options now are many, more complex, and certainly more costly. The other worlds still have the issues of *Silent Spring* to contend with because it will always be in the background. We hope the experience of the developed world will help less fortunate societies bypass the old problems.

Literature Cited

1. Graham, F., Jr. *The Dragon Hunters;* Dutton: New York, 1984.
2. Adkisson, P. L. *Bull. Entomol. Soc. Amer.* Fall 1986.

Appendix and Indexes

Appendix

Pesticide Names, Their Chemical Classes, and Their Principal Uses

This list is not intended to be all-inclusive. It includes only those compounds mentioned in the book or frequently mentioned elsewhere.

Trade or Common Name	Chemical Class	Principal Use
Acephate	Organophosphate	Insecticide
Alachlor	Chloroacetanilide	Herbicide
Aldicarb	Carbamate	Insecticide
Aldrin	Organochlorine, cyclodiene	Insecticide
Aminocarb	Carbamate	Insecticide
Atrazine	Triazine	Herbicide
Azinphos	Organophosphate	Insecticide
Benomyl	Benzimidazole	Fungicide
Calcium arsenate	Inorganic	Insecticide
Captan	Organochlorine	Fungicide
Carbaryl	Carbamate	Insecticide
Chlordane	Organochlorine, cyclodiene	Insecticide
Copper sulfate	Inorganic	Fungicide
2,4-D	Phenoxyalkanoic acid (organochlorine)	Herbicide
DDD	Organochlorine	Insecticide
DDE	Organochlorine	Is converted from DDT; DDE is not insecticidal
DDT	Organochlorine	Insecticide
DEF	Organophosphate	Defoliant (cotton)

0980-4/87/0203$06.00/0 © 1987 American Chemical Society

Dichlorvos	Organophosphate	Insecticide
Dieldrin	Organochlorine, cyclodiene	Insecticide
Difluron (Dimilin)	Benzoylphenylurea	Insecticide
Endosulfan	Organochlorine, cyclodiene	Insecticide
Endrin	Organochlorine, cyclodiene	Insecticide
Fenac	Organochlorine	Herbicide
Glyphosate	Organophosphate	Herbicide
Heptachlor	Organochlorine, cyclodiene	Insecticide
Kepone (chlordecone)	Organochlorine	Insecticide
Kelthane	Organochlorine	Miticide
Lead arsenate	Inorganic	Insecticide
Leptophos	Organophosphate	Insecticide
Lindane	Organochlorine	Insecticide
Malathion	Organophosphate	Insecticide
Methoxychlor	Organochlorine	Insecticide
Mevinphos	Organophosphate	Insecticide
Mirex	Organochlorine	Insecticide
Paraquat	Bipyridilium	Herbicide
Parathion	Organophosphate	Insecticide
Pentac	Organochlorine	Miticide
Permethrin	Pyrethroid	Insecticide
Perthane	Organochlorine	Insecticide
Phosalone	Organophosphate	Insecticide
Propanil	Acylanilide	Herbicide
Pydrin	Pyrethroid	Insecticide
Rotenone	Botanical, heterocyclic	Insecticide
Sodium arsenite	Inorganic	Herbicide
2,4,5-T	Phenoxyalkanoic acid (organochlorine)	Herbicide
TEPA	Organophosphate	Chemosterilant
TEPP	Organophosphate	Insecticide
Toxaphene	Cyclodiene	Insecticide
Trichlorfon	Organophosphate	Insecticide
Trifluralin	Dinitroaniline	Herbicide

Affiliation Index

Subject Index

A

Copy editing and production by Meg Marshall and Janet S. Dodd
Indexing by Keith B. Belton
Book design by Pamela Lewis
Jacket design by Carla L. Clemens

Typeset by Hot Type Ltd., Washington, DC
Printed and bound by Maple Press Company, York, PA

St. Louis Community College
at Meramec
Library